AF886482

Foundation Model v.1

Zeoform Value Transformation

How to Make Products from Plants using a Simple Nine-step Process.

A Model for Sustainable Manufacturing using Advanced Cellulose Technology.

Zeoform™ C/- Intertype
Unit 45, 125 Highbury Road
BURWOOD VIC 3125
www.intertype.com.au

Ordering Information:
Quantity sales. Special discounts are available on quantity purchases by corporations, associations, and others. For details, contact the "Special Sales Department" at the address above.

Zeoform™ Foundation Model v.1—1st ed.
ISBN 978-0-6453822-0-4

Foreword

Message from the CEO

We help businesses and entrepreneurs take the courageous step towards making sustainable products by supporting each of you to become leaders in the field of sustainability.

Put simply, your goal is the same as ours: to make high-value sustainable products. We do this through Zeo IP's high-level training, licencing and innovation services. Using proven systems and powerful frameworks of sustainable manufacturing, we help you to unleash the locked potential in your business.

We guide you from where you may be feeling a level of uncertainty or you're feeling overwhelmed with the amount of "green wash" in the world or perhaps you haven't been able to build the sustainable business of your dreams.

The strategies within this book have been used by businesses from around the world to change their value proposition forever using Zeo IP's technology and knowledge frameworks. Without the information included in this book, you will very likely miss the opportunity of a lifetime to transform your business using the most sustainable manufacturing material on the planet - plant fibre.

Choosing the right training, technology and making everything fit into your budget can be a really difficult process when you don't have a blueprint. My wish for you is that you use the knowledge in this book to gain confidence and clarity to make this life-changing decision without wasting your time and money.

Our clients experience these transformations every single day when we guide them through this process. That's why I'm so excited to share this book with you ... to help you accelerate your sustainability journey and create a business that contributes to the lives of others.

Alf Wheeler

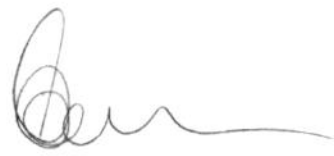

"We cannot solve the socio-economic and environmental problems with the same thinking we used when we created them."

- Zeoform

Contents

03	**Foreword**	
09	**Chapter 1**	Zeoform Value Transformation Method
15	**Chapter 2**	Pre-Production Process
27	**Chapter 3**	Active-Production Process
37	**Chapter 4**	Post-Production Process
43	**Chapter 5**	Zeoform Value Transformation Method
47	**Chapter 6**	Empowering the Pre-Production Process using ZMP
65	**Chapter 7**	Empowering the Active-Production Process using ZMP
83	**Chapter 8**	Empowering the Post-Production Process using ZMP
95	**Chapter 9**	Zeoform Value Transformation Applications
99	**Conclusion**	
102	**Next Steps**	
106	**Appendix A:**	Zeoform Value Transformation Applications
108	**Appendix B:**	Sustainability Self Assessment Guide

Acknowledgements

In response to emergent sustainability and manufacturing challenges in the construction, furniture and adjacent manufacturing industries, Zeo IP Pty Ltd developed these resources to meet the need for renewable material innovation.

Zeo IP Pty Ltd is a research and development company that specialises in renewable materials, and works closely with industry to support quality non-accredited training to meet future industry needs.

These resources were prepared by subject-matter experts in their chosen fields including Matthew Champion, Jesse Bird, Alf Wheeler and Martin Ernegg, and were developed as a collaborative effort to support the distribution of Zeo IP Pty Ltd intellectual property and proprietary processes.

Zeoform Value Transformation Method

Overview

Chapter Overview

Transformation Method

Why are some materials sustainable while others are not?

Industry representatives and governing bodies throughout the world have identified sustainability, innovation and the "common good" as a key priority to enable tangible outcomes that are good for economies, humanity and the planet.

A strong light is being shone on unsustainable business practices, and growing environmental concern is driving global organisations to reach for higher levels of ethical practice at unprecedented levels.

The root cause of the failure by businesses to prioritise the common good is not simply inadequate knowledge or poor education and training. Success is not to be found in excellent business management alone, or even the best levels of strategic planning, cooperation and solution innovation. The secret to successful sustainability is beyond the management activities that define good business practices. Successful sustainability, at its core, is rooted in something much simpler: How to turn abundant organic materials into high-value products without harming people or nature.

The Zeoform Transformation Formula

The Zeoform Sustainability Formula = CV x (ER + CG)

Commercial Viability (CV)
Environmental Responsibility (ER)
Common Good (CG)

The Zeoform Value Transformation Method

The Zeoform Value Transformation Method presented in this book is a framework for understanding sustainability at the community level. This model shows how you can take cellulose, the world's most abundant organic polymer, and transform it into high-value consumer products without having to rely on fossil-fuel composites and derivatives that are proving increasingly harmful. This Method then extends to show how businesses, governments and communities can increase the likelihood that their visions for circular economic success are implemented successfully.

The Zeoform Value Transformation Method has three key objectives. It's useful to think of these objectives as stepping stones to new opportunities for business and environmental success. All three elements must be in place for sustainable manufacturing to be realised.

Zeoform Sustainability Formula

Common Good represents the human dimensions of all connections and relationships between people and groups involved.

Commercial Viability represents an understanding of the nature of the business opportunity, why the opportunity is arising, and the risk of not taking up the opportunity.

Environmental Responsibility represents the physical dimensions and natural environment impacts of activities through the opportunity.

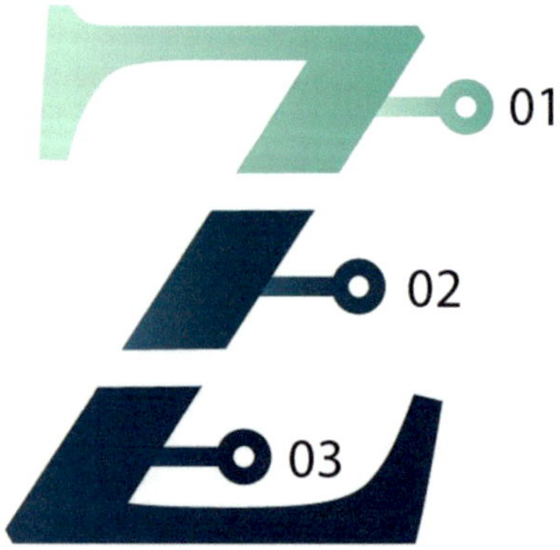

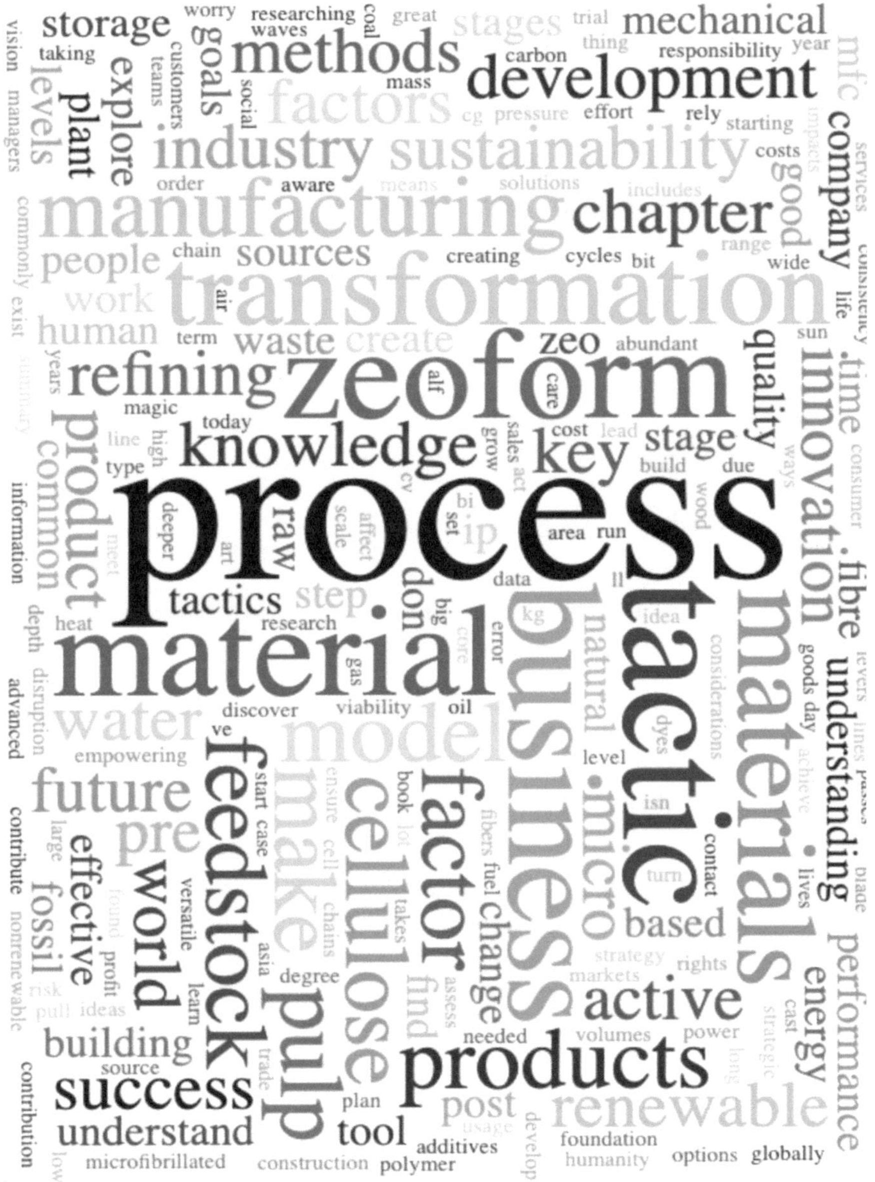
process
transformation
zeoform
material
manufacturing
tactic
business
materials
development
methods
sustainability
industry
factors
chapter
products
feedstock
pulp
cellulose
factor
model
make
knowledge
refining
innovation
key
mechanical
storage
success
understand
renewable
active
world
future
water
product
common
understanding
performance
energy
micro
fibre
company
quality
stage
natural
tactics
waste
sources
people
human
work
explore
plant
levels
goals
building
effective
fossil
change
based
tool
post
pre
create
step
raw
zeo
time
good

The Zeoform Sustainability Formula

The Three Pillars of Sustainable Development

Commercial Viability

Commercial Viability represents an understanding of the nature of the business opportunity, why the opportunity is arising, and the risk of not embracing the opportunity. Commercial viability also includes information about the internal and external drivers that created the opportunity, as well as the definition of returns on investment and effort.

Environmental Responsibility

Environmental Responsibility represents the physical dimensions and natural environment impacts of activities through the opportunity. Responsibility means taking care and ownership for all things, including for all pre-production processes, active process and post-production activities.

Common Good

Common Good represents the human dimensions of all connections and relationships between people and groups involved. Community contribution also includes the levels of cooperation, collaboration and coordination required, along with support for human rights and higher states of human relationships and development.

Summary + Knowledge Checklist

Chapter 01: Transformation Method

Tactic 1	The Zeoform Value Transformation Method helps organizations adopt sustainable manufacturing practices and opportunities.
Tactic 2	Sustainable manufacturing opportunities must meet three (3) key objectives: commercial viability, environment sustainability and contributing to the "common good".
Tactic 3	There are three (3) stages in the Zeoform Value Transformation Method: (1) Production pre-process (2) production active process, and (3) production post-process. For each of the three (3) stages of the Zeoform Value Transformation Method there are three (3) modules of know-how.

STAGES	KNOWLEDGE STEPS
Stage 1. Pre-Production process	1. Foundation knowledge 2. Sourcing feedstocks 3. Feedstock preparation
Stage 2. Active-Production process	1. Material transformation 2. Refining parameters 3. Additives and performance enhancers
Stage 1. Post-Production process	1. Water evaporation 2. Storage and shipping 3. Product innovation

Zeoform
Pre-Production Process

Overview

Pre-Production Process

Transformation Method

Chapter Overview

The first step to enabling a sustainable manufacturing system is to create an effective pre-production process,. This represents all activities that help you to acquire the knowledge, assess the opportunity and source the materials, equipment and feedstocks you need before you engage in manufacturing.

When new technology comes into the market, most people are naturally suspicious. New knowledge or technology represents a shift away from the generally accepted ways of doing things. Even when it brings with it the promise of new opportunities, its adoption doesn't follow a straight line as most people think.

This is particularly the case given the amount of volatility and change that is being experienced right now at the global level. Never before has humanity experienced mass-scale disruption to wealth distribution, supply chains and economic markets. During times of volatility, businesses begin to ask themselves questions and already feel some resistance to the changes underway.

Disruption is forcing the world to alter its course. Whether it's the COVID-19 pandemic, climate change, pressure on natural resources, political and social disruption - there will be winners and losers moving forward. Conservatism is still the mainstay of the dominant culture but the early adopters of innovation will leave the pack behind as they catch the early waves of change that run into the new world.

But change comes slowly at times. The mindset of humanity is a bit like a supertanker cargo ship - even when it is wanting to change course, the sheer weight and mass of the supertanker drives it forward in a straight line. It takes time for the ship to start coming about, and the same way one cannot steer a ship not in motion since there is no pressure on the rudder, one cannot drive change where there is no need, incentive or directive to adapt. During times of large scale change, it is those businesses who adapt the fastest that will stay ahead of the pack. Those that don't adopt new innovation will fall behind. McKinsey and Co. research show those businesses that invest in innovation during times of high market volatility have a 30% advantage over their competitors. But to take advantage, they must first become aware of the opportunity.

Building awareness of the need for new technology requires the following components to be addressed:

- What is the nature of future trends and how does it align with the vision, roadmap and strategies for the organisation?
- What is the risk of being left behind when a sudden and seismic shift in the market takes place, either through a regulatory, consumer or informational change that directly disrupts the market?
- How will the change impact your organisation, suppliers and community?
- What is the overarching motivation you have to grow and become more as an organisation?

Factors for Success in the Pre-production Process

In order to create an effective pre-production process, there are a number of factors that influence the way your organisation prepares for and makes good decisions, that are good for you but also for the environment and your community:

Factor 1:	An organisation's awareness of the global trends.
Factor 2:	How an organisation aligns to the UN Sustainable Development Goals.
Factor 3:	Depth of knowledge of fossil fuel consumption and waste accumulation.
Factor 4:	Depth of knowledge of sustainable material alternatives.
Factor 5:	Awareness of the potential of cellulose.

Each of these factors directly influences the success of creating an effective pre-production process.

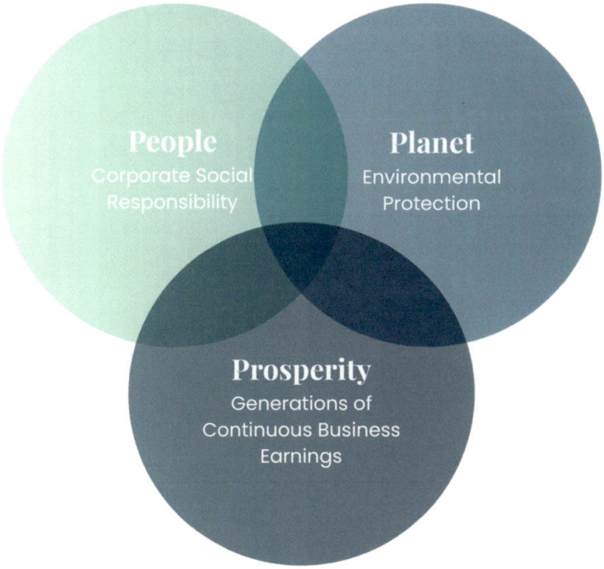

Pre-Production Process

Factors for Success

Factor 1: An organisation's awareness of the global trends

With so many trends and changes happening on a global scale, it's important that organisations are aware of these shifts to stay ahead. It can be difficult for companies to keep up with all the latest developments as they happen, but if you know what consumer groups care about most and where opportunities lie, then you can plan your strategy accordingly.

One of the best ways to stay on top of these trends is by reading industry publications, blogs, and other content that can help you understand how your customers think and behave. The more aware you are of current consumer needs and wants, the better prepared you will be to capitalize on opportunities as they arise. Do any of these insights ring true with your own experiences? What do others in your company know about this topic? Does it change the way things have been done historically? How could we leverage this information going forward?

LUCF - Land-use Change and Forestry

Figure Above UN Sustainable Development Goals

Factor 2: How an organization aligns to the UN Sustainable Development Goals

With the United Nations Sustainable Development Goals setting a global agenda for sustainable development, it is more important than ever that organisations have an aligned sustainability strategy. Companies need to be able to demonstrate their commitment by investing in developing new products and services with reduced environmental impacts, while also understanding how they can help build commercially viable and environmentally responsible business opportunities.

The sustainable development goals are the international agenda for reducing poverty, inequality and climate change. With these global challenges in mind, it's more important than ever that organisations align their activities to this agenda.

For instance, your company can develop products or services that help people live healthier lives without harming the environment or create jobs with a low carbon footprint. What steps has your organisation taken towards alignment?

Figure Previous Page: Top 10 Emitters. World Resource Institute, Science and Engineering Indicators. (2018).

Figure Above: UN Sustainable Development Goals

Factor 3: Depth of Knowledge of Fossil Fuel Consumption and Waste Accumulation

Researchers are making strides in the development of sustainable materials and products that will help reduce fossil fuel consumption, waste accumulation, and other environmental issues. The key to solving these problems is innovation.

If you want to contribute your part to this solution - find a way to make new materials or invent new processes for manufacturing existing ones. Your contribution could be as simple as changing how you consume resources by opting for regenerative materials over unsustainable materials.

You should be aware that there are many ways to reduce your impact on the environment. The first thing is to educate yourself about how much fossil fuel and other resources it takes to manufacture a product or service. This will help you understand what sustainable alternatives exist for existing products and services as well as new inventions in development.

One way to do this would be by using free sustainability calculator tools which can help you calculate an estimate of resource usage based on inputs from various sources such as industry average values or input from interviews with subject-matter experts. With these numbers in hand, it becomes easier to make decisions about which materials and processes should be adopted at different stages of production.

The current market size of the U.S. construction sector is valued at around $1.36 trillion U.S. dollars by 2030 with global construction opportunities expected to rise to almost $12+ Trillion.

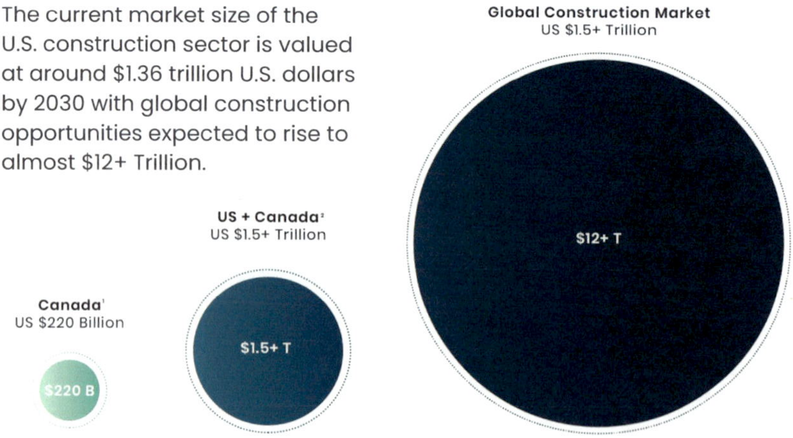

Global fossil fuel consumption

Global primary energy consumption by fossil fuel source, measured in terawatt-hours (TWh).

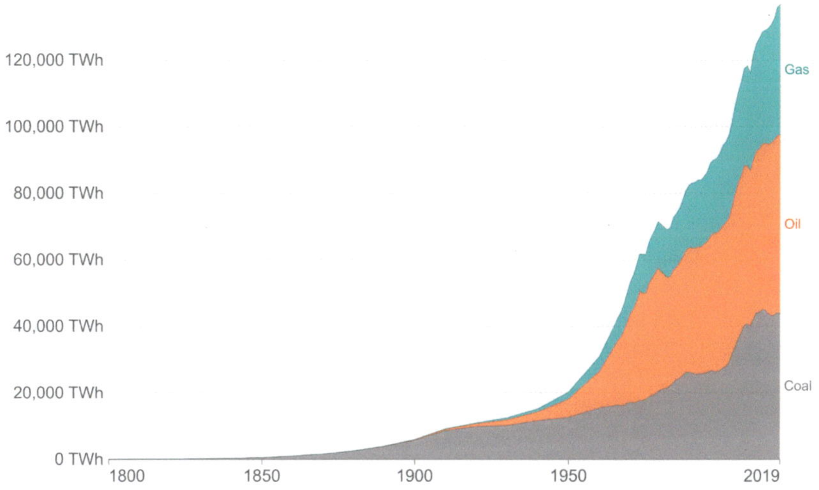

Factor 4: Knowledge of Sustainable Material Alternatives

Traditional materials can be substituted with renewable, sustainable ones. The potential of new technology and innovation is to create a more eco-friendly future for our planet by creating new products that do not rely on unsustainable material alternatives.

Learn about some other innovative companies working toward a better world. The construction industry (along with many others) faces a huge challenge in the coming decades. With more and more people moving into cities, we will need to find new ways to build up rather than out if we want to preserve green space and protect our environment. As many of us have been hearing so much about lately, this change is going to require innovation.

If you are looking for help identifying materials that use renewable, sustainable material alternatives or solutions using innovative technology, contact Zeo IP today! We can provide expert guidance on how best to work with these changing times while continuing your growth as a company. What other companies do you know that are taking steps towards sustainability? Work together with them to create a formidable force by building on each others' greatest strengths and biggest opportunities.

Factor 5: Awareness of the Potential of Cellulose and Cellulose Feedstocks

The benefits of cellulose-based materials are clear, and they can be used to make a wide variety of products. From textiles like clothing or carpets, to paper for writing with pens on the go, there's no limit to what you can do with these renewable resources!

If you want help finding out more about how organic polymers such as cellulose-based wood replacements and biopolymers can replace petrochemical synthetics and toxins, and wood and wood-composites in your company's supply chain then contact us today. We offer consultation services that will evaluate your current needs and suggest solutions that could provide long term value.

Cellulose is not only the most abundant organic polymer on earth, but it's also one of the most versatile. This material can be found in everything from paper to trees to plant-based polymers and even agricultural waste! This means that second generation materials like cardboard and industry bi-products can be leveraged to make you money sustainably while contributing to the circular and regenerative economy.

So if you're looking for an alternative fibre source with water resistant properties and you don't want to add oil-derived phenol or formaldehyde resins — look for a fibre that still contains lignin - a natural component in plants that has unique hydrophobic properties.

Primary Feedstocks

Hemp Cellulose

Industrial hemp is one of the most available and widely produced bast fibres with high cellulose content. Interest in these fibres is warranted due to environmental protection challenges as well as their inherent properties such as low density, high specific strength, and stiffness.

Bagasse Fibres

Bagasse, also called megass, fibre remaining after the extraction of the sugar-bearing juice from sugarcane. Bagasse is the essential ingredient for the production of pressed building board, acoustical tile, and other construction materials and can be made into a number of biodegradable plastics.

Bamboo Pulp

Bamboo has received increasing attention over the last two decades for its economic and environmental values. In Africa, Asia and Latin America, it is closely associated with indigenous culture and knowledge and is widely used for housing, forestry, agroforestry, agricultural activities and utensils. It's even a food!

Summary + Knowledge Checklist

Chapter 02: Transformation Method

Tactic 1 The first step to enable a sustainable business is to plan for and implement an effective pre-production process.

Tactic 2 Become aware of the macro and micro trends and market drivers affecting the future of your customers and organisation.

Tactic 3 An organisation's alignment with the United Nations Sustainable Development Goals enables them to contribute to a thriving planet and the common human good.

Tactic 4 Organisations need to be fully informed of the degree of fossil-fuel consumption globally, deforestation, and the impact of the overconsumption of plastic and other synthetic polymer products on the environment.

Tactic 5 Uninformed organisations are inclined to make decisions without considering the environment.

Tactic 6 Increased awareness of material alternatives that are sustainable and renewable empowers an organisation by giving them choices to move away from unsustainable practices.

Tactic 7 Cellulose is the most abundant and renewable source of organic polymer on the planet. When an organisation learns how to "tap into" the potential of highly-refined cellulose they enter into a new and wondrous world of sustainable production.

ZEOFORM™

Zeoform Active-Production Process

Overview

Active-Production Process

Transformation Method

Chapter Overview

The active production process is the second stage of the Zeoform Value Transformation Method and represents how you can make effective manufacturing materials from various sources of pre-pulped cellulose.

> *Enabling you to transform manufacturing with renewable materials, utilise additives to optimise performance, and future-proof your business.*

Economic dependence on non-renewable materials has dramatically changed since the first industrial revolution. Approximately 80 percent of the total amount of energy used globally each year comes from fossil fuels. But a major problem with fossil fuels, aside from their being in limited supply, is that burning them releases carbon dioxide and methane into the atmosphere. Rising levels of heat-trapping carbon dioxide and methane in the atmosphere is the main cause of global warming.

Natural resources like cement and stone (that make concrete) and wood are considered by some as infinite resources but these materials can not be relied upon forever. Major resources like oil, portland cement, and other metals like lead, zinc, and copper will be gone within 50 years if we don't change our usage habits now. It is reported that a chain reaction can occur if one material is depleted from the resource chain. This is why banning plastic would create an unprecedented large-scale market disruption because there isn't another readily available replacement for its uses yet.

Another problem is that many industries and consumers don't acknowledge what happens with products after their service life. Often ending up in landfills, incinerators, or piling up in high-volume storage zones awaiting a responsible solution.

Looking around, you can argue that the most commonly used building materials in construction today consists of concrete and steel. Unlike wood however, concrete is made through unsustainable practices. Wood can be torn down to be reused, and concrete, which is part of a down-cycling system, has a limited use to become aggregate in other concrete applications.

Types of non-renewable resources

There are four major types of non-renewable resources: oil, natural gas, coal, and nuclear energy. Oil, natural gas, and coal are collectively called fossil fuels. They are found in underground layers of rock and sediment. Plants, algae, plankton and other early life bloomed in ancient times, absorbing sunlight and storing energy through photosynthesis. When they died, the organisms drifted to the bottom of the sea or lake. Pressure and heat worked together to transform the plant and animal remains into crude oil (refined into petroleum), coal, and natural gas.

We use oil for many things. About half of the world's petroleum is converted into gasoline. The rest can be processed and used in liquid products (petroleum) such as nail polish and rubbing alcohol, or solid products such as water pipes, building products, insulation, roofing, and thousands of other items. In many cases the knowledge for implementing an active process is clear. For example, there are standard industry practices within the Pulp and Paper Industry that are already used to create abundant sources of clean, washed, de-inked and pre-pulped sources of recovered waste paper and old corrugated cardboard. There is also readily available pre-pulped cellulose that is sourced from bamboo, hemp, wheat straw, and sugarcane bagasse amongst others.

Several factors, as shown below, will impact the successful achievement of the active process of the Zeoform Value Transformation Method.

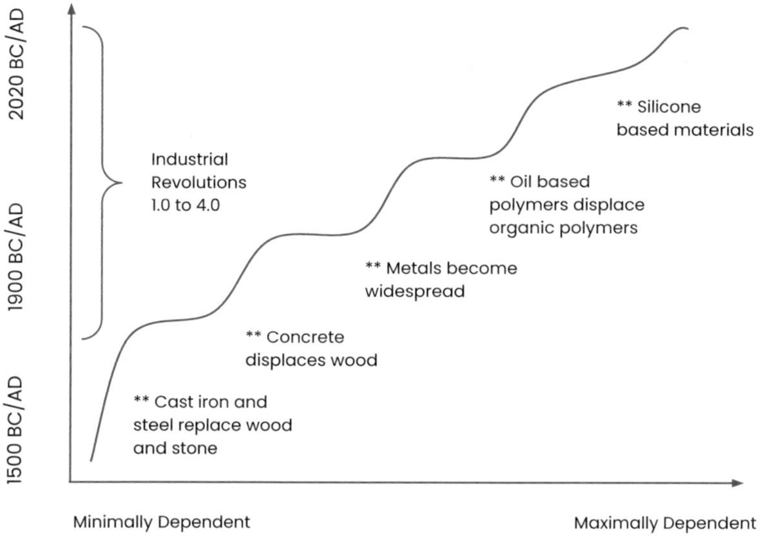

Factors for Success in the Active-production Process

Factor 1 Effective material preparation.

Factor 2 Proven process and reliability.

Factor 3 Versatility of the process.

Factor 4 Scalability of the process.

Factor 5 Optimisation of performance.

Each of these factors directly influences the success of creating an effective pre-production process.

Humanity's dependence on non-renewable materials has had a staggering rise since the advent of the industrial revolution.

Active-Production Process

Factors for Success

Factor 1: Material Preparation

When it comes to sourcing sustainable materials, efficiency must be balanced with strategic intent. This means that you need to look at the best options for both your company and for the environment when deciding where and how to source raw goods. It is not enough just to use a new innovation or technology if it does not result in any improvement over what was already being done.

Paper is a perfect example of a material that participates in the circular economy as it remains one of the most recycled products in the world, creating significant environmental benefits and raw material resources to many industries. Indeed, maintaining a consistent level of quality whilst limiting the environmental footprint of the product has become a major challenge for the industry. Our proprietary formula can be structured around this abundant resource and help you become an industry leader, by providing you with the higher than usual added-value in the repurposing of waste materials.

Sourcing sustainable materials is not an easy task. It requires an investment in time and patience, but it's worth the effort in the long run to protect our environment for future generations.

Factor 2: Process Reliability

In the pursuit of success, it is easy to get lost in all of the different techniques and methods out there. It can be hard to know which ones are reliable or produce consistent results. The best thing we can do is experiment with a variety of options while always looking for proven systems that have been tested by others before us.

Focus on how to use proven techniques and methods. These will produce consistent and reliable results when executed correctly. For example, we know from neuroscience research that people are more likely to purchase products if they can visualize themselves using them in their own life scenarios. This is why the best marketing videos include a before-and-after story of someone who used the product so well for such an extended period of time that they achieved some sort of success or lifestyle improvement as a result. Be willing to experiment with trial and error but always look for systems that have been shown through empirical evidence (i.e., data) to work consistently over time instead of relying on hopeful strategies.

Factor 3: Versatility of the Process

The world will never stand still, so you might as well embrace it and allow your business to grow alongside the changing needs of the market. You can do this by being open-minded about new opportunities that pop up or adapting to a change in customer preferences - all while staying ahead of what's coming next for your industry. Stay adaptable to these ever-changing elements of success and innovation.

You may be thinking that adapting your company's strategy will take too much time or money; but if you want market disruption as an opportunity for growth, then do not hesitate! See change and market disruption as an opportunity to grow and evolve into new eras of success led by the changing elements of the market.

Don't be shy to partner with innovative businesses that have done the hard work for you. Let them tailor their products and services for businesses like yours who are looking for a competitive edge over their competition — all while staying ahead of the pack as you walk side-by-side into the new world of sustainability and business ethics.

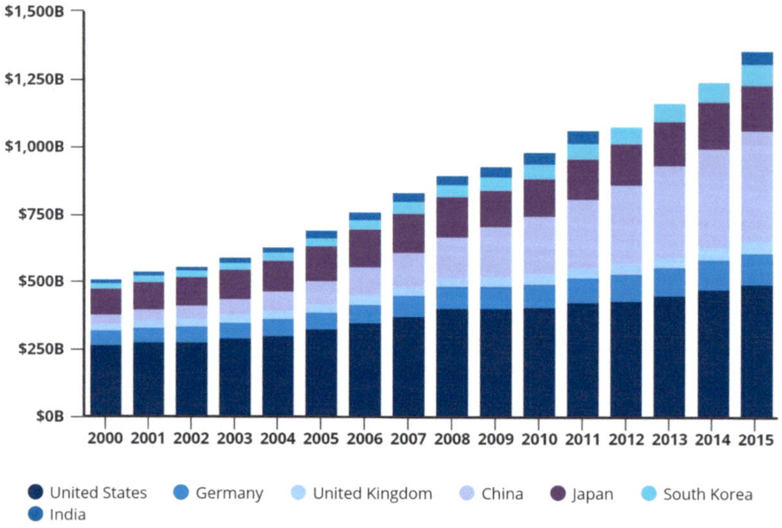

(Source: National Science Board, Science and Engineering Indicators. (2018). Global R&D: One Measure of Commitment to Innovation.

Factor 4: Scalability of the Process

The world is always looking for the next big thing, but what does it take to develop something that can actually make an impact?

Great innovations are only truly effective if they have a big impact and are used widely. Developing a breakthrough idea or innovation is one thing, but scaling it to make a contribution to the lives of others is another. Consider how you will work with your team members at every step of the development process so you're able to scale into success quickly. You may want to set up milestones or deadlines for yourself as well as objectives within each milestone.

Build a vision in which you scale into next levels of success and production. Whether that means making incremental improvements on what exists today, innovating at the edges with new ideas for products and services, or inventing entirely new things altogether, there are many ways to turn your big idea into something even bigger than yourself!

Factor 5: Optimisation of Performance

Success is not a destination, it's an ongoing journey. The art to success is to consistently look to improve and grow. Continued growth and learning are the key ingredients for long-term future success. One of our most important roles of innovation-based businesses is helping others to develop their production process so they can produce goods more efficiently than ever before while also optimising opportunities that will keep them competitive with other companies in the region or around the world.

Knowledge Insight

Production active-process is the second stage of the Zeoform Value Transformation Method. This is an essential step for the opportunity to be realised, for the organisation, nature and the community as a whole. Understanding how to manufacture sustainably by making Zeoform Micro Pulp can be a simple process in some cases and a transformation in thinking in others.

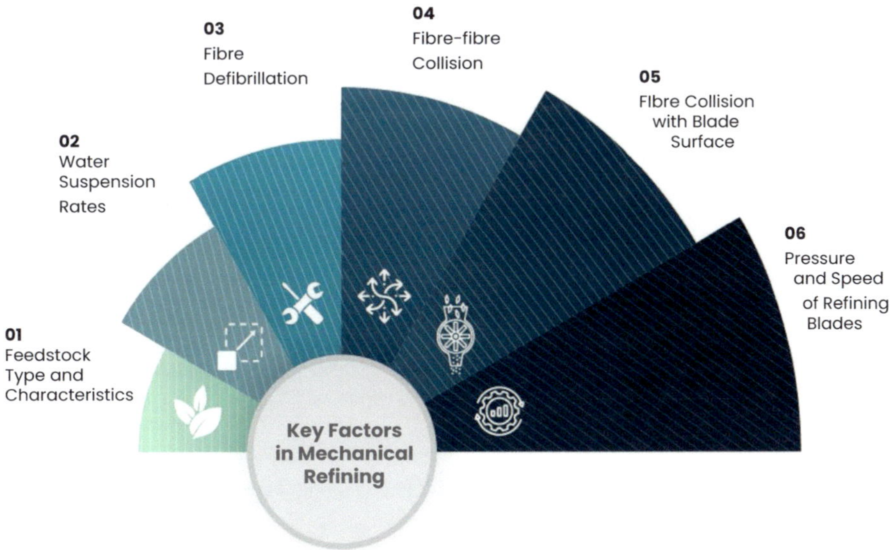

Summary + Knowledge Checklist

Chapter 03: Active-Production Process

Tactic 1 The second step to enable a successful and sustainable business is to create an effective active-process.

Tactic 2 Humanity needs to dramatically reduce its dependence on non-renewable resources.

Tactic 3 Sourcing renewable materials is not an easy task. Put in the time, energy and effort and you will reap the benefits.

Tactic 4 Build a vision in which you scale into next levels of success and production. Maintain routine feedback loops with employees and partners.

Tactic 5 Maintaining precise control of the active process is critical to producing consistent results. This will provide you with accurate projection models and reliable sustainable growth.

Tactic 6 The versatility of the active process to adapt to changing needs.

Tactic 7 The opportunity for the active process to scale in line with growth.

Tactic 8 Optimisation through the use of performance additives and enhancers.

ZEOFORM™

Zeoform Post-Production Process

Overview

Post-Production Process

Transformation Method

Chapter Overview

The post-production process is the third and final stage of the Zeoform Value Transformation Method. The post-production process includes any action or event that strengthens the value of the offering and increases the opportunity to contribute to a better world.

The post-production process does not always require major elements or events. Several factors, as shown below, will impact the successful achievement of the active-process of the Zeoform Value Transformation Method.

Factor 1: Efficiency of material storage and stock control.
Factor 2: Effectiveness of marketing and sales.
Factor 3: Creating next generation product innovation.
Factor 4: Financial and operational management.
Factor 5: The power of positive partnerships.

Each of these factors directly influences the success of creating an effective production post-process.

Post-Production Process

Transformation Method

Factor 1: Efficiency of Material Storage and Stock Control

In order to have an efficient and profitable business, you need the right amount of raw materials on hand. That's where stock control comes in - it is used by every organisation that uses raw materials and finished products. It applies at every stage of production from purchase and delivery all the way through using and reordering stocks.

Stock control ensures that capital is not tied up unnecessarily, protects your productions if there are problems with your supply chain and helps ensure quality assurance as well as customer satisfaction by having a continuous flow of goods available for sale or service.

Factor 2: Efficiency of Marketing and Sales

Marketing effectiveness is the cornerstone of every company's goals. Marketing and sales are not just about getting people to buy your products. The key is making sure that they keep coming back for more, which means you need to develop a marketing strategy which focuses on the customer while also increasing brand visibility. Ensure that you analyse what makes customers tick at different stages in the purchasing process or lead funnel. Be sure to implement these techniques before creating any new digital strategies!

You will always win the day if your marketing continually lowers the costs of finding and winning business, while also increasing the value of that business. The key to success comes down to understanding what drives customers' purchasing decisions today - at both ends of their buying journey. Whether you're a small business owner, solopreneur or an entrepreneur - your marketing and sales strategy needs to be effective. If it isn't, you risk not seeing the growth that's necessary to make your business thrive.

Factor 3: Creativity and Success in Product Innovation

Innovation is the process of consciously utilising new ideas, or finding new uses for existing ideas, to create new value in response to a need. It's an important distinction that for any idea or concept to be considered truly innovative, it must lead to the creation of new value. Therefore, an innovative idea must meaningfully create measurable or definable social, technical, organisational, economic or environmental value.

When innovation is applied to developing new products, it can lead to new applications for your existing product portfolio. And, when it's applied also to business decisions - from product development to organisational restructuring - the results can be remarkable.

Contrary to popular belief, innovation is a creative process that is highly structured and repeated. When done well, it gives you the structure, purpose and accountability you need to create new ideas and products in response to a changing world.

Innovation Cycle

Innovation is the creation of an effective new result in that something is improved or its value has increased. This can include how your business generates revenue and profit, develops new products and services, serves customers, staff, shareholders, society and the environment.

Factor 4: Profit Levers

There are a number of levers that businesses can pull when they want to improve their performance. For manufacturers this includes reducing the cost of goods sold, reducing days payable, or reducing inventory days turnover. Remember that each lever has an impact on your profitability so don't forget to optimise your business.

The levers that you pull to increase profitability should be based on the needs of your business. You may find it useful to look at each lever and decide which ones are best suited for your situation. The factors mentioned above make up a majority of what you see in most businesses' income statements.

Factor 5: The Power of Positive Partnerships

The world we live in is changing — and the way we create partnerships should change with it. A successful partnership agenda requires inclusive, equal and trust-based connections — at the global, regional, national and local levels. Positive partnerships are built upon shared principles and values, and upon an inclusive vision that puts people, planet and profit alongside each other.

Strong levels of cooperation are needed now more than ever to ensure that businesses have the means to recover from pandemic, build back better, and achieve common good based goals which are great for the planet.

Gone are the days of working alone; businesses that are prepared to cluster together and leverage a pool of collective strengths and opportunities will succeed faster than those that don't.

Summary + Knowledge Checklist

Chapter 04: Transformation Method

- **Tactic 1** — Efficiency of material storage and stock control.
- **Tactic 2** — Effectiveness of marketing and sales.
- **Tactic 3** — Creativity and success in product innovation.
- **Tactic 4** — Financial and operational management.
- **Tactic 5** — The power of positive partnerships.
- **Tactic 6** — The opportunity for the active process to scale in line with growth.

- **Summary** — The post-production process is not always the most glamorous or exciting part of a company's production cycle. However, what happens in this stage has a significant impact on how successful your product will ultimately be.

  By carefully examining how each of the factors listed above affects your post-production efforts, you will be able to identify areas that need improvement and then come up with strategies to fix them. These are just six important considerations but there are many more, so take some time out today to look at where you can make changes in order to increase efficiency!

ZEOFORM™

Zeoform Value Transformation

Overview

Zeoform Value Transformation

Transformation Method

The Zeoform Value Transformation Method has three (3) core stages that define the basic building blocks for successful manufacturing:

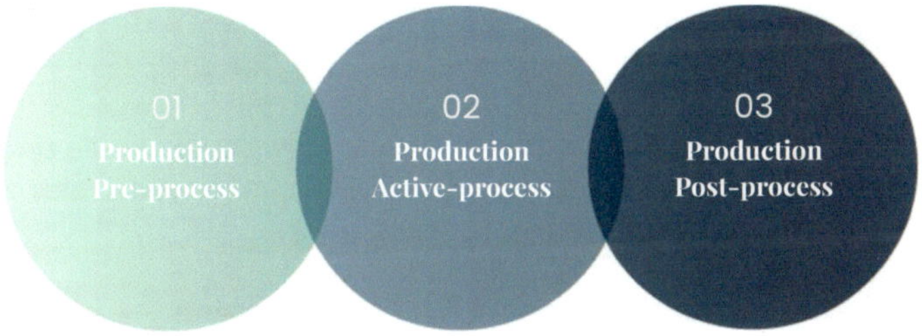

Zeoform Value Transformation is a breakthrough technology that represents a market opportunity in the sustainable materials space. Zeoform Value Transformation can be used:

- As a learning tool for teams considering innovation technology in the sustainability sector.
- As a discussion tool to support your organization when thinking about innovation and new technology.
- To guide your growth efforts when seeking new business opportunities that are commercially viable, environmentally responsible and contribute to a better community.
- As a coaching tool for managers and teams.
- To diagnose your growth projects to assess where blockages and misalignment might occur.

Sustainable manufacturing is realised when all three goals are achieved.

1. Commercial viability
2. Environmental responsibility
3. Common human good.

First the business must be commercially viable in order to stay in business and meet its objectives and obligations to its owners. Next, the business and all of its owners and directors have a moral obligation to maintain a healthy and vibrant natural world for all present and future generations of human and nonhuman species. Last, the business must contribute positively to the common human good. That is, it must adhere to the universal rights and freedoms of all human beings - without discrimitation or limitation.

The building block nature of the Method is important in terms of application of the Zeoform Value Transformation for business, government and community. For entrepreneurs and business managers, the Zeoform Value Transformation Method is simple and easy to apply. When an element early in the Method is weak, then sustainability and success begins to break down.

When sustainable manufacturing is not being achieved, the Zeoform Value Transformation Method provides a framework that is simple and easy to follow. Using this framework you can analyse your sustainable manufacturing plan, assess your strengths and identify new opportunities for growth.

ZMP Stages	**Key Factors of Driving Success**
Stage 1: Pre-production process	Factor 1 - Awareness of the global trends. Factor 2 - Alignment with the UN Development Goals. Factor 3 - Knowledge of fossil fuels and waste accumulation. Factor 4 - Depth of knowledge of sustainable alternatives. Factor 5 - Degree of motivation for growth and advantage.
Stage 2: Active-production process	Factor 1 - Material preparation. Factor 2 - Process reliability. Factor 3 - Versatility of the process. Factor 4 - Scalability of the process. Factor 5 - Optimisation of performance.
Stage 3: Post-production process	Factor 1 - Storage and stock controls. Factor 2 - Marketing and sales. Factor 3 - Next generation product innovation. Factor 4 - Profit levers. Factor 5 - Positive partnerships.

ZEOFORM™

Empowering the Pre-Production Process using ZMP

Overview

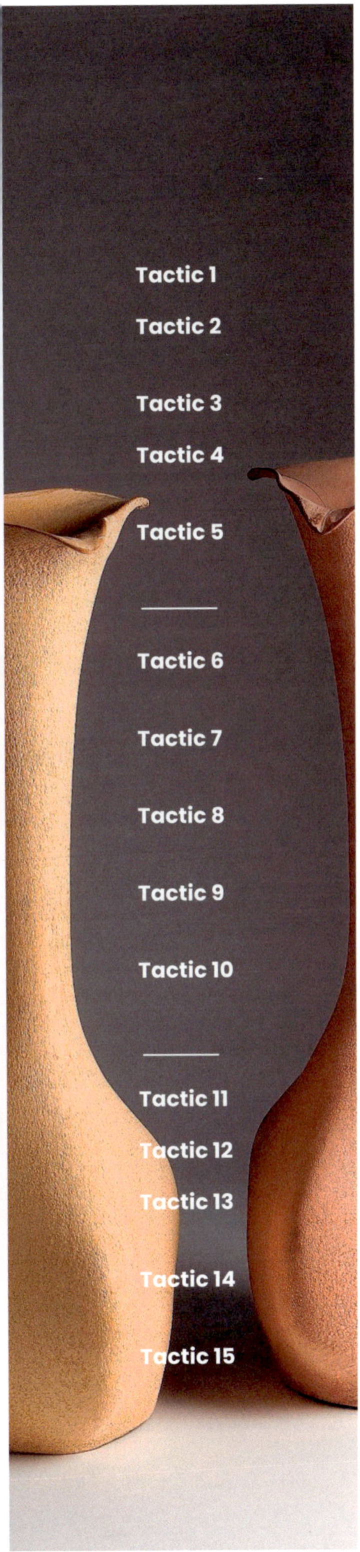

Chapter 6 Overview

Tactic 1	Gain a deeper knowledge of cellulose.
Tactic 2	Understand more about plant fibre morphology.
Tactic 3	Explore the future of manufacturing.
Tactic 4	Awareness of methods for nano and microfibrillated cellulose (NMFC) production.
Tactic 5	Understand the basis of mechanical refining to make NMFC.
Tactic 6	Discover the abundance of cellulosic biomass.
Tactic 7	Identify commonly available feedstocks in your area.
Tactic 8	Explore all considerations in how to make a feedstock viable.
Tactic 9	Explore globally popular feedstocks such as hemp, bagasse and bamboo.
Tactic 10	Understand the importance of identifying feedstock substitutes.
Tactic 11	Gain knowledge of feedstock planning
Tactic 12	Feedstock volumes and quality
Tactic 13	Look at preparation methods such as washing and pre-pulping
Tactic 14	Identify opportunities for material performance enhancement
Tactic 15	Look at other factors that affect the production process.

Chapter 2 Review

Recall from Chapter 2 the factors that influence the degree to which you are strategically prepared for sustainability, including:

- An organisation's awareness of the global trends.
- How an organisation aligns to the Sustainable Development Goals.
- Depth of knowledge of fossil fuel consumption and waste accumulation.
- Depth of knowledge of sustainable material alternatives.
- Degree of motivation for growth.

Because of these factors, the act of planning and preparation does not always produce the best results. For example, employees of organisations may not be well enough informed about a business's need to change towards being more sustainable. Even if all employees are provided with the same information, some may discredit aspects of it whilst others may have a strong will and conviction to meet all three goals: commercial viability, environmental responsibility, and common human good.

The pre-production process is an important part in every successful manufacturing business. Sustainability is not just achieved based on the research that is conducted, but rather how the research is applied and actioned by the business.

Using Zeoform material innovations to make business more sustainable, let's explore the specific tactics of the Zeoform Value Transformation Method to build foundational knowledge and create effective pre-production processes that leverage cellulosic fibre as the sustainable material.

Empowering the Pre-production Process

Before these sustainable manufacturing tactics can be applied, it is necessary to discuss and agree the goals you want to achieve. The key decision makers in the business must have a common and well developed understanding of:

- How sustainable manufacturing aligns with the vision of the business.
- The level of commitment the business has to becoming a sustainable business; and why this is strategically important in the present planning horizon.
- The risks associated with not changing
- The consumer and regulatory environment changes, competitor threats and Industry 4.0 developments that will affect the industry.
- When key decisions need to be made.
- The constraints affecting the business from developing and growing.
- The resistances in the culture of the business environment stopping the business from moving forward.

Once the foundation knowledge has been established, a combination of the additional strategies and tactics can be used to advance the sustainability of the business. Let's look a bit deeper into each of the above tactics.

Tactic 1: Efficiency of Material Storage and Stock Control

Cellulose is a natural polymer (made up of long chains of microscopic compounds that are linked together) and one of the most abundant natural compounds on earth. It's mostly found in plants but it's also found in recycled material and even bacteria — in fact it makes up all plant cell walls and is commonly found alongside lignin and hemi-cellulose chains. But cellulose isn't just an ingredient; it has many different applications for things like medicine, construction materials and textiles. A deeper understanding of cellulosic materials can help you find new uses for these versatile substances or improve the ones that already exist!

Learn more about cellulosic materials, what makes them so special, and how you can use them to manufacture your own products. Zeoform has perfected the art of refining cellulose into a structural, mouldable material for making boards, sheets and shapes of all kinds of densities. The more educated we all become about these wonderful plant-based products, the better off our planet will be!

Tactic 2: Understand more about Plant Fibre Morphology

Plant fibre morphology is the study of plant anatomy that looks at the size, shape and arrangement of plant fibres. Understanding how these morphological features develop in plants can give us an idea about what type of fibre combinations are possible for different materials and applications. For example, flexible fibres like cotton or silk may be better suited to make paper because they are not as brittle. While hemp cellulose and lignin may be better utilised for structural applications. By combining various lengths and fibre characteristics you can make a fibre "matrix" that is perfect for various applications. For example, Zeoform Micro Pulp is made from a novel blend of fibre morphologies that together combine to create its versatility and strength.

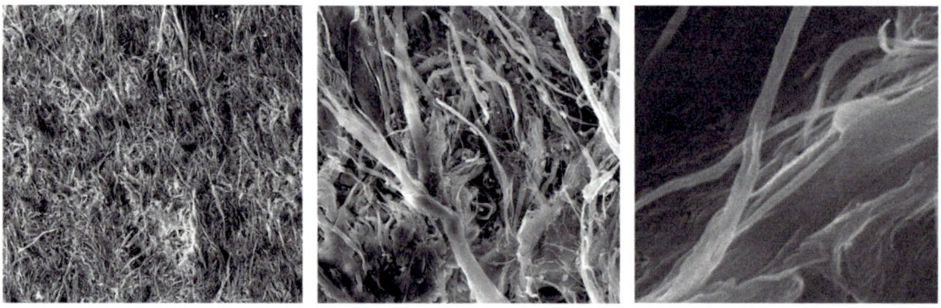

View Field: 1.5mm *View Field: 150um* *View Field: 3um*

Figure. SEM Images of Zeoform Micro-pulp under 200x magnification. ZeoLabs Canada. 2020

Tactic 3: Explore the Future of Manufacturing

The fusion of Industry 4.0 developments together with sustainable development goals will heavily influence the future of manufacturing. This fourth industrial revolution known as Industry 4.0 refers to an era in which manufacturers are increasing their use and incorporation of robotics into production lines along with smart manufacturing technology all while incorporating modern connectivity methods such as cloud computing and big data analytics.

As your business navigates this new landscape, align yourself now so you can be ready for tomorrow's opportunities by building a plan that includes digital transformation, new materials, robotics and sustainable designs. The benefits of this are increased automation and production line efficiency, more accurate data for decision making, decreased lead times on product delivery, better quality control over products being manufactured and higher levels of transparency into supply chains.

Align your business with this fourth industrial revolution by incorporating Industry 4.0 innovations to help you meet sustainability goals while still increasing productivity and contributing to a thriving planet. The following page provides a brief overview of industrial developments and their impact on renewable materials since the year 1760.

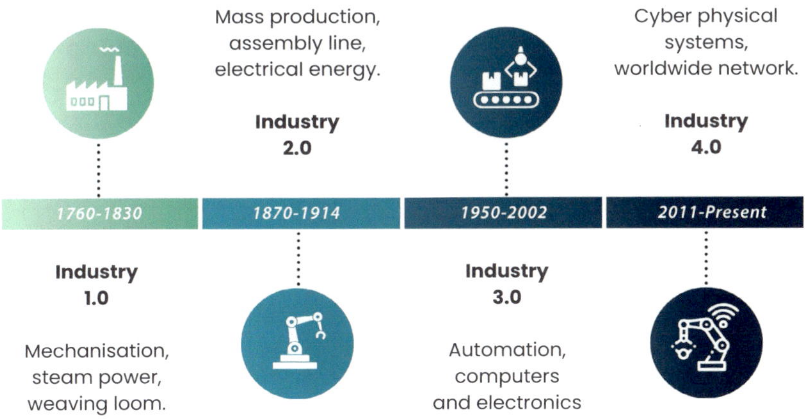

Industry 1.0

Mechanisation (1760 - 1830)... machines and tools replace animal and human labour; widespread use of iron and steel for machinery. During this era, cast iron and steel displaced wood and stone in building structures.

Industry 2.0

Mass Production (1870 - 1914)... assembly lines and automations increase efficiency and productivity; widespread use of stainless steel and plastics. During this era, concrete replaced wood in large structures and aluminium replaced wood in lightweight structures.

Industry 3.0

Digitisation (1950 - 2002)... electronics and information technology is integrated into manufacturing and communications; widespread robotic automations and Programmable Logic Controllers (PLCs) help robots do their job at very specific times and points in the production process. During this era, metal became the dominant material of engineering and oil-based polymers replaced natural fibres, pottery and wood.

Industry 4.0

Interconnectivity (2011 - present)... higher level automation driven by artificial intelligence and optimised manufacturing using real-time data and sensors; integration of cyber-physical systems (machine to machine) across the supply chain. During this era, silicon-based communication controls all commerce and social activities.

> There is much debate as to what the next industrial revolution will be; some suggest that biomimicry will drive the development of new consumer and manufacturing configurations; regardless, due to resource constraints it is believed that all developments will require a transformation at the individual and collective levels of human consciousness. That is, humanity will need to return to pre-industrial dependence levels on non-renewable materials and this will require a systemic shift in values and belief systems.

Tactic 4: Awareness of Methods for Nano and Microfibrillated Cellulose (NMFC) Production

Nano and microfibrillated cellulose (NMFC) is a new material that has many benefits for industrial and commercial use. It can be created in multiple ways using different techniques to achieve similar results. It is a raw material with unique properties due to its unique length-weight ratio of less than 30 nm wide and almost 120 nm long with a tensile modulus of over 130 GPa bringing them close to Kevlar® fibres. While this is true, mechanical properties of fibrillated cellulose mixes are complex, and are best evaluated with methodology relevant to their application.

As such, it's not surprising why this versatile product continues to grow as an alternative to traditional and non-renewable building material across various industries. However, there are still challenges faced by those who wish to create NMFC products because access may vary depending on location and availability of equipment needed for production methods. We recommend you gain awareness of methods for nano and microfibrillated cellulose (NMFC) production so you are able to choose the most cost-effective way to produce this versatile product!

01 Enzyme Assisted Preparation

02 Tempo-oxidising (chemical)

03 Steam Explosion Treatment

04 High-shear mechanical refining

05 Microwave (ultrasonic) Treatment

06 Homogenisation

*or a combination of enzyme hydrolysis, shear and homogenisation.

Tactic 5: Understand the Basics of Mechanical Refining to make NMFC

Mechanical refining of cellulosic material is one way to ensure a consistent, high quality product. Zeo IP uses mechanical refiners to make Zeoform Micro Pulp and they've been refining for over 10 years. Our process is dependant on the following: **pulp/fibre morphology, mechanical treatment**, and **specific grinding parameters.** (feedstock, energy, and grinding)

Knowing these three key elements can help you understand how our manufacturing system works as well as what it does for your business! We'll show you the key elements in the process of making a special blend of cellulosic material suited for a large variety of manufacturing purposes.

Mechanical refining of cellulose is a method that has been used extensively over many decades because it offers significant advantages related to product quality and availability in comparison to other methods such as acid or alkaline processes. Traditional techniques often use alkaline processes which only cover a limited range of products due to their chemical nature. Herein, mechanical refining can produce high-quality materials from all types of cellulose-based feedstocks.

Tactic 6: Discover the Abundance of Cellulosic Biomass

Cellulosic biomass is a renewable resource that can be found in all ecosystems on Earth. Biomass also supports life through its production of oxygen during photosynthesis - without biomass there would not have been enough time before natural extinction events started wiping out plant species during geological history timescale.

We can help you find the right sources of biomass for manufacturing NMFC from a wide selection of cellulosic biomass materials that come from sustainable sources like wood chips, agricultural waste, food production waste, recycled paper and textiles, and so on.

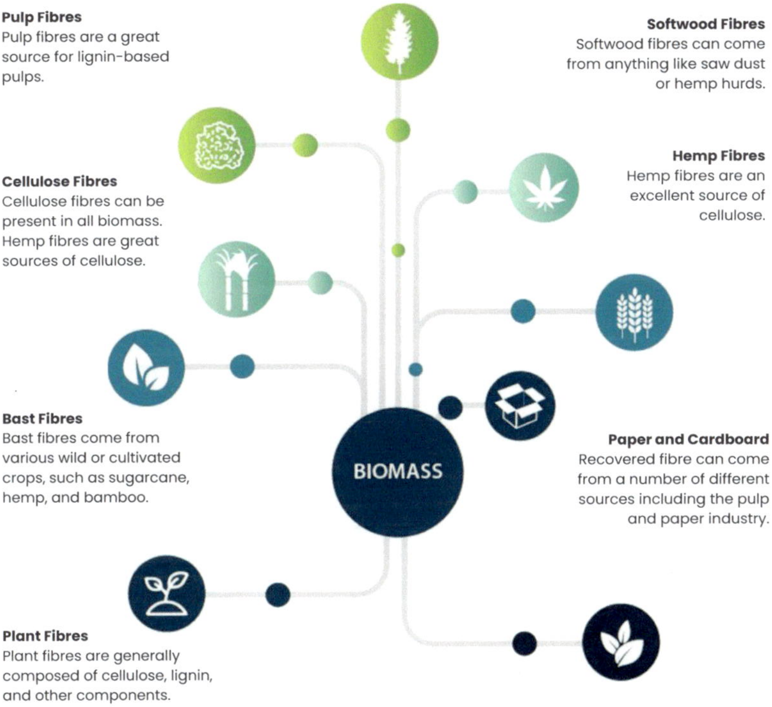

Figure Above: Top 9 Biomass feedstocks that may be available to you. Each feedstock is unique and should be considered prior to investment.

Tactic 7: Identify Commonly Available Cellulosic Feedstocks in your area

Explore the different array of feedstocks that are available in your area and how they may be used. The more you know about these resources, the better prepared you will be to find a sustainable solution for generating renewable materials from biomass.

If there is one thing we want you to take away from this book it is that cellulose comes in many forms which vary by region - so understanding what feedstock options work best with your geography can help make decisions on which materials make sense for you. Some of these feedstock sources you may be familiar with and others might not have been on your radar until now. We encourage you to identify a few cellulose feedstocks to try in your area so you can start exploring how they work for yourself!

Figure Above: Developing connections with local resources is key to sustainable manufacturing. It is important to know both major and minor supply chains in your region and the factors that will impact its efficiency.

Tactic 8: Explore all Considerations that make a Feedstock Viable

You'll need a reliable and affordable feedstock to produce your products. Researching the most viable sources of this feedstock will help you reduce risks in production, which can lead to consistency in product quality.

Make sure that you explore all considerations for making your chosen source of feedstock workable before jumping into any agreements or contracts. It may be worth exploring other options if it turns out that they are not as lucrative or cost-effective as initially thought because changing course is much easier now than after investing significant time and money upfront.

We recommend considering not just price when looking at different sources, but also how reliable suppliers are with delivering goods time after time without delay or interruption, as well as the quality of what they deliver based on industry standards.

Figure: Material Alternatives. Hemp Harvesting shown in grey.

Tactic 9: Explore the Benefits of a Globally Available Feedstock such as hemp, bagasse and bamboo

Zeoform Micro Pulp is a sustainable, renewable and biodegradable product that can be made from hemp, bagasse or bamboo pulp. These feedstocks are versatile for use in various applications including benchtops, shapes, and insulation boards. Hemp offers the most versatility of all three with its ability to produce fibres that can be used in many different types of products.

Figure: Material Alternatives. Bagasse Fibre Harvesting shown in grey.

However bagasse and bamboo is available in large volumes in a diverse range of geographies. Hemp, bagasse and bamboo are three examples of commonly available renewable sources that can be used as feedstock for our system. Explore the benefits of hemp, bagasse and bamboo before deciding which one will best meet your needs. Recovered paper is also commonly available and versatile so you'll never need to worry about running out of materials.

Tactic 10: Understand the Importance of Identifying Feedstock Substitutes

With so much uncertainty in the world, it's important to be prepared for any disruption that could derail your supply chain. We explore how to identify feedstock substitutes that future-proof production from climate, trade or other disruptions with proven strategies.

As the world's population grows and global warming continues to disrupt natural cycles, it is more important than ever for manufacturers in every industry to have a plan of action when disruptions occur. One key strategy that can help you future-proof your production from climate, trade or other disruptions is identifying feedstock substitutes. If your company uses hemp to manufacture products and suddenly there are no more hemp suppliers available due to climatic events or any other reason, what will happen? Develop proven strategies to protect yourself from this type of disruption so you never find yourself in an emergency situation without options.

Figure: Material Alternatives. Waste paper is a great source of cellulose and lignin,

Feedstock Alternatives

Old Corrugated Cardboard (OCC)

Consumer packaging and tissue will be driven largely by demographic shifts and consumer trends such as the demand for convenience and sustainability. We expect innovation to be a critical success factor, particularly in light of recent concerns over plastic packaging waste, which could harbour both opportunities and challenges for fibre-based consumer packaging.

Recovered Waste Paper (RWP)

Newsprint, coated papers, and uncoated papers are great alternatives for lignin and cellulose based feedstocks. They are often processed through heavy 'down cycling' systems but through Zeoform's revolutionary technology, we can help you transform your supply chains to *make money sustainably.*

Softwood

The majority of biomass carbon is stored by plants in forests as wood. What's even more interesting is that 30% of the world's fibre supply is derived from softwood. The recovery of the disposed goods from this market is a great contender for Zeoform's Transformation Method.

Tactic 11: Gain Knowledge of Feedstock Planning

Materials planning is a critical component for any production process. Improperly managing the feedstocks required by a refinery can have significant impacts on profitability, resource allocation and more. Gain knowledge of this important process to ensure your facility continues running smoothly.

The desired characteristics and quality must be determined along with ensuring there are no shortages or overages in stock levels, which has significant implications on profitability, allocating resources effectively, etc. If you're not sure where to start when it comes to feedstock planning for your company, don't worry; experienced consultants can help!

Tactic 12: Feedstocks + Volumes

In order to produce a higher yield of bio-based products, manufacturers should ensure that they have enough biomass for refining. This means having an ample supply of feedstock in terms of volume and quality. There are important considerations when considering these two factors so you can be more efficient with your plant based materials.

Feedstock volumes are typically measured by weight per unit area (kg/m3) or weight per hectare (t/ha). Quality is an important consideration because it affects how quickly and easily the material can be processed to produce high value results, as well as its suitability for use in different applications.

Tactic 13: Look at Preparation Methods such as Washing and Pre-Pulping

The preparation process of feedstock can be very different depending on the type. If you're not sure how to prepare your particular kind, consult with a professional before proceeding any further. This will help ensure that all of the equipment is operating at peak performance for maximum efficiency and productivity. It also helps avoid damage or wear-and-tear on both machinery and product as well as reduce safety risks during production processes.

The type and quality of your feedstock will depend on what you are refining, but there are some steps that apply to all! Some preparation may be required, like washing and pre-pulping, before feeding your raw material into the refiner. Remember these preparation methods when preparing for production so you can be confident that everything goes smoothly with no surprises!

Tactic 14: Identify Opportunities for Material Performance Enhancement

The use of natural dyes and pigments is a common practice in textile production. From ancient times to the present day, people have been using colours extracted from plants, minerals or animal sources for purposes ranging from ritualistic decoration to artistic expression. Discover some of the more common colouring and performance enhancement methods used when producing Zeoform Micro Pulp that you might find interesting if your focus is on novel and sustainable business practices.

If you identify opportunities for material performance enhancements during any stage of your supply chain process, don't hesitate to innovate! Cutting-edge solutions are often the result of trial and error, and a few false starts. This is the nature of innovation and the growth process.

Tactic 15: Look at other Factors that affect the Production Process

Choosing the right feedstock for your production process can be confusing, but it's important to consider all of the factors that affect how a particular feedstock will behave. If you are interested in whole plants as an option, make sure they have been adequately washed and prepared before refining so that any contaminants or foreign substances don't get introduced into your final product.

We recommend looking at other considerations like water quality, volumes and feedstock storage because this could change what equipment is needed throughout the process.

Additionally, it may not make sense to use only one type of feedstock if you have access to more than one kind - using different types will allow you to produce various end products with distinct properties that could serve multiple purposes in industrial applications. Before making any final decisions about what would work best for your business needs, look at all factors affecting the production process so you know how they might affect your bottom line, the environment and all generations of humanity - present and future.

Tactics Summary

Pre-production process is an opportunity for you to make great decisions and investments. When multiple sustainable manufacturing tactics are applied, a process results where:

- Decisions become well informed.
- There exists a common understanding among your decision makers and key stakeholders.
- Planning is made that de-risks not just your investment but also future generations of products that are more sustainable, relying less on non-renewable materials.

Summary + Knowledge Checklist

Chapter 06: Transformation Method

- **Tactic 1** Gain a deeper knowledge of cellulose.
- **Tactic 2** Understand more about the plant fibre morphology.
- **Tactic 3** Explore the future of manufacturing.
- **Tactic 4** Awareness of methods for nano microfibrillated cellulose (NMFC) production.
- **Tactic 5** Understand the basis of mechanical refining to make NMFC.
- **Tactic 6** Discover the abundance of cellulosic biomass.
- **Tactic 7** Identify commonly available feedstocks in your area.
- **Tactic 8** Explore all considerations in how to make a Feedstock viable.
- **Tactic 9** Explore globally popular feedstocks such as hemp, bagasse and bamboo.
- **Tactic 10** Understand the importance of identifying feedstock substitutes.
- **Tactic 11** Gain knowledge of feedstock planning
- **Tactic 12** Feedstock volumes and quality.
- **Tactic 13** Look at preparation methods such as washing and pre-pulping
- **Tactic 14** Identify opportunities for material performance enhancement

ZEOFORM™

Empowering the Active-Production Process using ZMP

Overview

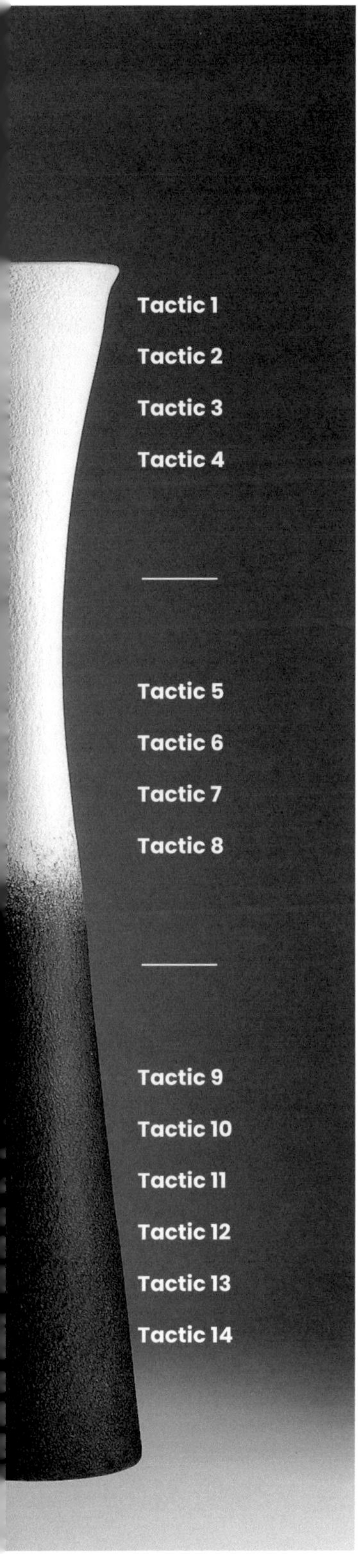

Chapter 7 Overview

Tactic 1 The magic of cellulose

Tactic 2 Nature's transformation process

Tactic 3 The 3-Stages of cellulose transformation

Tactic 4 Benefits of Zeoform Micro Pulp

Now we've looked at material Transformation, let's look at how to improve performance of the materials.

Tactic 5 Material characterisation

Tactic 6 Material additives

Tactic 7 Pigments and dyes

Tactic 8 Materials applications

Finally, let's look at the mechanical refining tactics to make Zeoform Micro Pulp.

Tactic 9 Researching mechanical refiner types

Tactic 10 Researching refining blade setups

Tactic 11 Understanding refiner cycles and passes

Tactic 12 Gaining knowledge of energy consumption

Tactic 13 Learning about fibre & water ratios

Tactic 14 Other key variables

Zeoform Value Transformation

Transformation Method

Chapter 3 Review

Recall from Chapter 3 the factors that influence the degree to which you understand and are strategically prepared for sustainability, including:

- Raw material preparation.
- Process reliability.
- Versatility of the process.
- Scalability of the process.
- Optimisation of performance.

As a basic principle, businesses must first view the task of developing a plan that delivers on their strategic goals, meets customer desires and achieves social, nature and human rights goals.

This chapter examines the sustainable manufacturing tactics of using Zeoform Micro Pulp to influence the above factors for creating an effective production active process. Starting with a range of tactics to establish refining parameters:

Empowering the Active-production Process

Transformation Method

Tactic 1: The Magic of Cellulose

The magic of cellulose is that it can be broken down into its cellular components and repurposed into nano-and micro-fibrillated componentry. This means that everything from paper to textiles, wood products, and more can be reused many times over without ever going anywhere near a landfill or incinerator.

With our help you might even find new ways for your company to use this miraculous material! Have we piqued your interest in the power behind cellulose? If so, feel free to contact us about how we could work together with you on developing a sustainable business model. We'd love to hear what creative ideas come out of such an endeavour!

Another incredible advantage of cellulose is its ability to be recycled back into the process to form new usable materials. This can be done in the production line or after a products in-service life. All these items are made from the same basic material but there's a lot more than meets the eye when you look at how they break down or what they could become in their next life cycle!

If you want to learn more about this fascinating process for recycling old products so they don't end up in landfills, contact our team today!

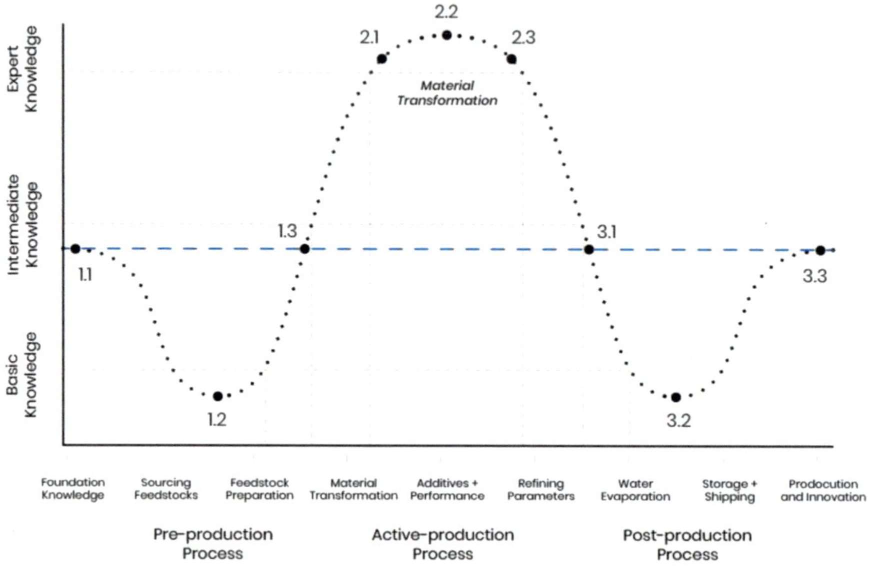

Tactic 2: The Transformation Process

Zeo IP has developed a process that transforms naturally occurring cellulose into nano-and micro-fibrillated cellulose materials. This technology is the backbone to how cellulose material binds and forms next level structures and bonds for use in manufacturing. To achieve these breakthroughs and outcomes, Zeo IP technology harnesses nature's 3-Step Method for transformation. These are:

Step 1: Current state - there is an existing form or structure in place.
Step 2: Active process - a proven step-by-step process occurs.
Step 3: Transformed state - a new and evolved physical form or structure is made.

Material innovation is dependent on new active-processes being created which can be used to transform existing materials into new forms and applications. At the heart of the Zeoform Value Transformation Method is an active process that creates Zeoform Micro Pulp.

Tactic 3: The 3-Stages of Cellulose Transformation

The active process at the heart of the Zeoform Value Transformation Method is a transformational process that turns cellulosic fibres into a "standard" fibre morphology made up of nano and microfibrillated cellulose (NMFC) called Zeoform Micro Pulp. The three stages of cellulose transformation are:

Pre-Production: Cellulose fibres
Cellulosic fibres are multi-component structures made from cellulose polymers, i.e. cellulose chains. The cellulose chains in the fibres are attached to each other to form elementary fibrils that allow for greater bonding points in the drying process. Several elementary fibrils are bound to each other to form microfibrils and several microfibrils form aggregates. The links between the cellulose chains, elementary and microfibrils are known as hydrogen bonds.

Active-Production: Zeoform Value Transformation Method
The manufacturing conditions used to convert macro-scale cellulose into its nano-scale form have a critical influence on the dimensions, composition, and properties of the resulting product. Zeo IP uses a mechanical refining process and equipment to achieve the novel Zeoform Micro Pulp that gives it its unique qualities and characteristics.

Post-Production: Zeoform Micro Pulp
The newly created nano and microfibrillated cellulose (NMFC) matrix is a material where the individual microfibrils have been partly or totally detached from each other. Better performance in terms of strength development is associated with a higher degree of fibrillation. However, it is reported that additional fibrillation does not contribute to further strength gains that can justify the increasing production cost. Higher fibrillation tends to lead to higher densities with less porosity, with additional flexible qualities.

1) Some images adapted from Honext Sustainable Solutions. 2019

2) Source: Bird, Jesse, "Building the Paper Economy: Finding cultural and spatial agency through recycled paper." (M.Arch Thesis., Azrieli School of Architecture & Urbanism, Carleton University, 2020) Carleton University Research Virtual Environment.

Tactic 4: Benefits of Zeoform Micro Pulp

Introducing the revolutionary material that changes everything. Made from cellulose fibres and water – and absolutely nothing else, our process technology transforms this natural resource into an industrial-strength moulding material capable of being formed into a diverse array of product applications!

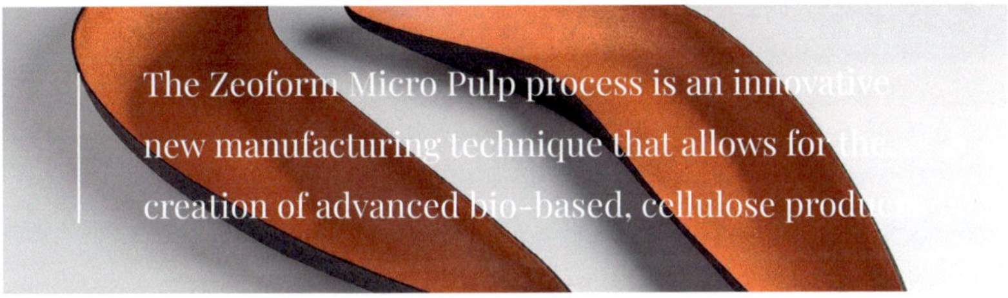

When you use this process to create your next product line with us, we can help you eliminate the use of unsustainable materials by using only plant fibres and CO2 sequestering technology. We also offer sustainability consulting services so if you're interested in boosting local economies or creating jobs through circular economy practices; contact us today!

Tactic 5: Material Characterisation

Zeoform Micro Pulp is a product with great potential - it has the ability to be customised and tailored for any type of product or need, as well as being sustainable and eco-friendly. Generally cellulose has good mechanical stability, however, the mechanical behaviour of Zeoform Micro Pulp is dependant upon the degree of refining which improves the degree of the fibrillation of the fibre. Factors affecting the fibrillation include the source of the cellulosic fibre, refining process including mechanical separation, pre-treatments, and the duration of hydration etc.

Tactic 6: Material Additives

There are a range of natural and synthetic additives that can play a defining factor in your product. We encourage you to go deeper into the basic and more exotic additives that can give you competitive advantage. At Zeoform we use organic micro-pulp additives to improve material properties, increase shelf life etc.

Whatever needs you have for your product, we can help you understand what may be worth considering when formulating your batch. With all these new insights into the world of additives at hand, there is no better time to start experimenting with some basic or even exotic ones!

Tactic 7: Materials Pigments and Dyes

Pigments are commonly used as a key component of paint, plastics and printing. They can be natural or synthetic in origin with some pigments being more stable than others. The Zeo IP team has covered the range of natural and other pigments that we use for proven results when it comes to embedding colour, product appeal and efficiency into Zeoform Micro Pulp.

If you're considering utilising pigment colours to create eye-catching products or if you need help sourcing them from suppliers such as ours, get in touch with us today! We have all sorts of great tips about how we use these materials to produce high quality products that stand out against competitors. It is important to consider what pigments you plan to use so that it allows the product to remain biodegradable and be recycled.

Tactic 8: Materials Applications

With all of these functionalities, we can see that NMFC is a versatile and exciting material for many different applications. If you're looking to incorporate this innovative material into your end products, you should consider your key requirements then assess to see how our micro pulps could enhance the function or look of your final product. Some application examples include:

- **Benchtops and pressed boards** - As a benchtop material, Zeoform Micro Pulp is an ideal material outright or as an additive, to make high-value, biodegradable pressed boards.

- **Moulded forms and technical shapes** - With its strength and versatility, Zeoform Micro Pulp can provide new lightweight, all natural materials for creating lighter and stronger products than most wood-composite manufacturers. This provides a competitive edge that will allow your company to grow sustainably and responsibly.

- **Insulation** - As an insulation material, Zeoform Micro Pulp can become a leader in the packaging and building industries with its thermal and strength functionalities. This gives ZMP the ability to replace some synthetically derived products like petroleum-based foams and boards, or provide reinforcement and stabilisation to foams, joints, adhesives, etc.

- **Additives in Hemp Blocks etc**. - The versatility of Zeoform Micro Pulp is a perfect natural additive for numerous products and industries. Whether you're in the fibreboard or concrete industries, your imagination is your limit for how you can use Zeoform Micro Pulp to give your products the edge you're looking for. Contact us today and find out how and where our material can be best utilised!

Figure: Zeoform Original Didgeridoo Instrument. 2018

Tactic 9: Mechanical Refiner Types

Mechanical refining is an important part of the process in making Zeoform Micro Pulp. If you're looking to refine your own Zeoform Micro Pulp, then understanding how mechanical refiners work will be helpful for designing a plan that works best for you and your company.

There are many different types of mechanical refiners and we've covered the main ones in this lesson, but there may be other options available for you depending on your specific needs. You should explore the different types of refiners, and refiner setups, that are available and how they can help make more viable products at a range of production levels and scales. Whether you are planning for an entry-level or industrial scale production, engineering solutions exist for all business needs.

Tactic 10: Refining Blade Setups

Zeoform Micro Pulp is made by feeding cellulosic feedstocks into the center of a disk refiner, consisting of two grooved discs. With one stationary disk and one in motion, the feedstock is forced toward the periphery by the centrifugal force and is sheared into pulp between the discs.

One of the constraints of the mechanical refining process is that it requires a larger amount of electrical power compared to other methods e.g. chemical methods. As the world shifts its energy generating processes away from non-renewable resources like oil and gas to viable renewable energy sources like solar and wind power, energy use becomes less relevant. An important part of developing an efficient refining process is managing the disk set-ups. The mechanical refining blades are made from large rotating circular disks with various grooves and patterns. Work with us and learn the many blade configurations and parameters that must be considered for optimum usage and output for your feedstock and product.

Refiner Considerations

There are a number of factors to consider when investing into a refiner including the refiner type, the blade set up and groove pattern, the number of passes and cycles, the pressure and speed of the disks, and of course your feedstock. Blade configurations play a crucial role in the degree of fibrillation you wish to achieve. Work with us to determine what is the best set up for your needs and we will put you in touch with our distributor!

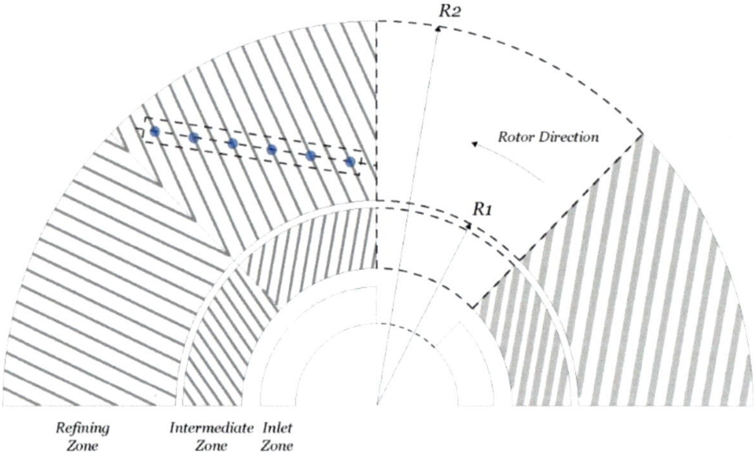

(b) *Cross-view*

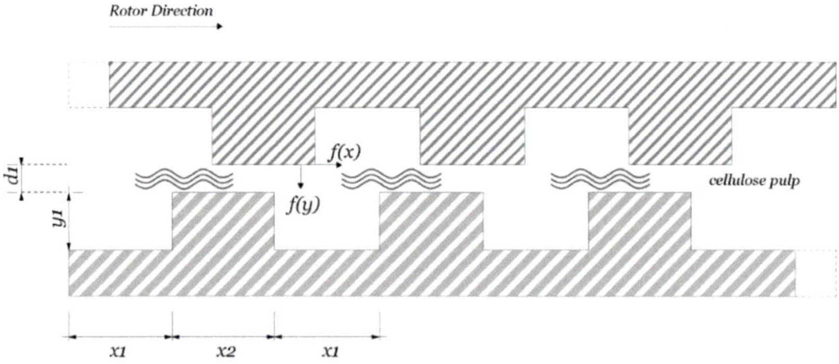

Tactic 11: Refiner Cycles and Passes

The number of refining cycles and passes varies depending on the type of pulp you are making. For example, softwood fibres take approx 20 refining cycles to produce an acceptable quality kraft paper, whereas hardwood fibre can take 30 or more refining cycles. Depending on your specific product needs, you may require less or more refining cycles and passes. Pre-pulped hemp cellulose usually takes about 10-15 cycles - it is our favourite feedstock because of its durability, abundance and overall mechanical properties. There are indicators to look out for at different stages in the process - these will help you determine how close your pulps are to being ready for refinement.

Make sure that all required parameters have been met so that when finished with processing they can be used for manufacturing!

It's important to have a good understanding of basic mechanical refining so you can be more knowledgeable when looking at your own fibre supply or trying to troubleshoot process problems. If there are any questions about anything else related to mechanical refining please don't hesitate to contact us!

Tactic 12: Energy Consumption

Efficient energy consumption is key to business success. For efficient and effective energy-conservation, it is essential to keep these rates as low as possible whilst optimising productivity output. There are some common averages along with other important performance indicators that will help you make your decision on how best to invest in your manufacturing process without exceeding budgets or carbon allowances.

But what are the benefits of keeping your power usage low? Well it could mean savings on production costs and greater efficiency which in turn leads to more profit margins. Indeed there are some average rates for how much electricity each kg requires (kWh/kg) along with other important metrics. Energy optimisation means choosing and adopting methods and policies for correct use of appropriately generated energy that is desirable from a national economical point of view and guarantee energy stability.

Tactic 13: Plant-fibre & Water Ratios

Cellulose is made up of thin fibres, or strands or fibrils, which are held together with hydrogen bonds. These strong connections are what gives wood it's ability to bend without breaking because of how sturdy its structure really is!

Zeoform Micro Pulp is made up of highly refined cellulose using a mechanical refining process which transforms these fibres into a cellulose matrix that can be used for a range of applications. The resulting Zeoform Micro Pulp is a pulp, water-based suspension with a fine network of microfibrils.

The fibre-to-water ratio in this water-based suspension is very important for the performance of Zeoform Micro Pulp. We have found the perfect balance for refining cellulose until it reaches what we call our "standard" specification. However, different material characteristics are possible but it is not dependent on the initial fibre to water ratio. Work with us to learn how to reduce transportation costs of nano and microfibrillated cellulose (NMFC) through additional pre and post-processing treatments.

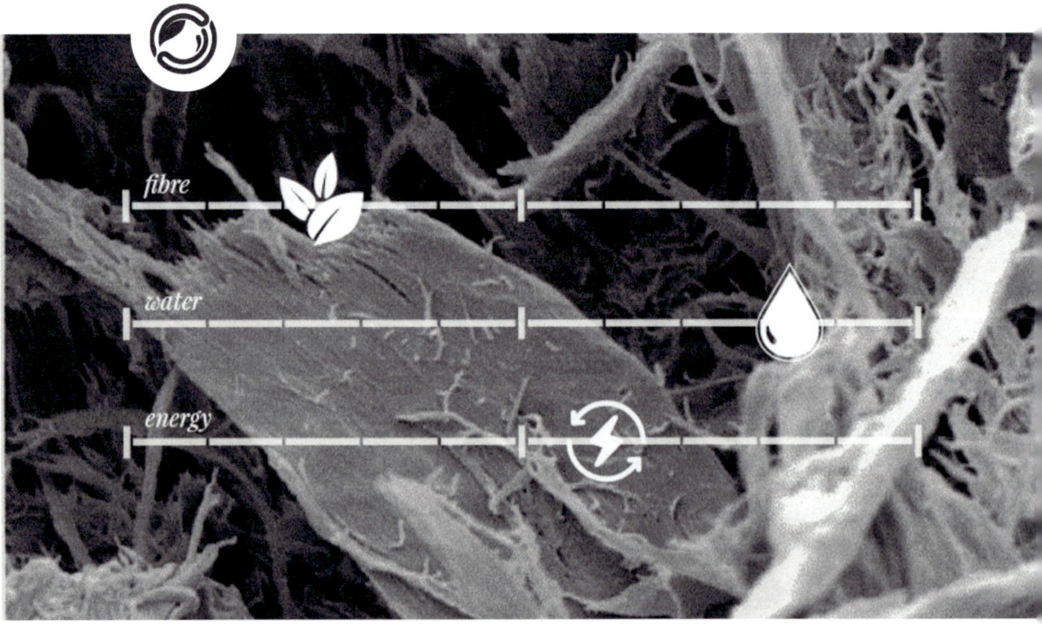

Tactic 14: Key Variables

Zeo IP makes Zeoform Micro Pulp using a mechanical refining process. There are a range of variables which must be considered and managed to create a consistent flow of high quality and efficiency. These variables include: the maintenance and function of a high quality refiner; preparation and cleanliness of the cellulosic feedstock; fibre to water consistency levels; disk refining blade setup; amount of energy consumed KwH/kg; number of material passess through the refiner; speed and rpms of the refiner during the different stages of refining.

When these key variables are managed and repeated, a high quality Zeoform Micro Pulp is produced. Of course, some of these variables can also be adjusted and alternated to create intentional variations to the characteristics of the product.

Tactic Summary

Multiple tactics can be used to help you succeed and become a sustainable manufacturer. The production active-process stage for making Zeoform Micro Pulp is your opportunity to implement technology and innovation aligned to a sustainable future and which will give you a competitive advantage over the market. You can achieve this by:

- Leveraging the magic of cellulose to manufacture sustainable products.
- Building an efficient and effective process to turn plant-fibre into high value renewable materials.
- Custom-designing Zeoform Micro Pulp to meet the exacting needs of your customers; including creating unique combinations of material qualities and characteristics.

The most successful sustainable manufacturing initiatives focus their effort on proactive steps that drive great outcomes. Next we will look at the post-active process for building sustainable businesses using Zeoform Micro Pulp.

Summary + Knowledge Checklist

Chapter 07: Empowering the Active-Process

Tactic 1	Understand the Magic of Cellulose
Tactic 2	The Zeoform Transformation Method
Tactic 3	The 3-stages of cellulose transformation
Tactic 4	Benefits of Zeoform Micro Pulp
Tactic 5	Material characterisation basics
Tactic 6	Material performance additives basics
Tactic 7	Basic understanding of pigments and dyes
Tactic 8	Materials applications and potential markets
Tactic 9	Researching mechanical refiner types
Tactic 10	Researching refiner blade setups and configurations
Tactic 11	Understanding refiner cycles and passes
Tactic 12	Gaining knowledge of energy consumption
Tactic 13	Learning about fibre & water ratios
Tactic 14	Key variables for how to empower the active production process

ZEOFORM™

Empowering the Post-Production Process using ZMP

Overview

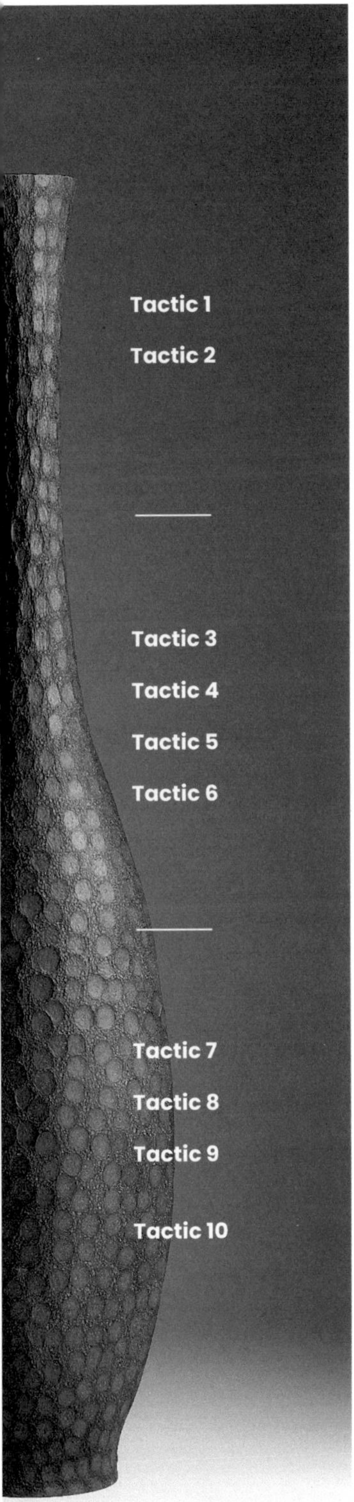

Chapter 8 Overview

Tactic 1 — Water evaporation fundamentals

Tactic 2 — Water evaporation basics and advanced methods

The next step, once you have established the water evaporation fundamentals is to look more closely at the storage and shipping process:

Tactic 3 — Material storage

Tactic 4 — Material stock control

Tactic 5 — Material shipping

Tactic 6 — Material quality assurance

Next you need to adequately consider business and product innovation parameters that can support longer term business success:

Tactic 7 — The production innovation process

Tactic 8 — The future of construction and furniture

Tactic 9 — Developments in wood replacements, and other adjacent markets

Tactic 10 — The research and development process

Zeoform Value Transformation

Transformation Method

Chapter 4 Review

Recall from Chapter 3 the factors that influence the degree to which you understand and are strategically prepared for sustainability, including:

- Materials storage and stock control.
- Marketing and sales.
- Product innovation.
- Profit levers.
- Partnerships.

As a basic principle, businesses must first view the task of developing a plan that delivers on their strategic goals, meets customer desires and achieves social, nature and human rights goals.

This chapter examines the sustainable manufacturing tactics of using Zeoform Micro Pulp to influence the above factors for creating an effective production active process. Starting with a range of tactics to establish refining parameters.

Empowering the Active-production Process

Factors for Success

Tactic 1: Water Evaporation Methods

The evaporation of liquid like water is the process by which a liquid turns into vapor or gas. It can occur directly from the surface of the material; it can also be carried out through pressing and heating to speed up the process. Understanding the basics of evaporation and how it applies when making Zeoform Micro Pulp is important.

The main element that can enhance or control evaporation is heat, so by using purpose built equipment and process design you are able to speed up this process and achieve maximum efficiency out of your material!

Tactic 2: Water Evaporation Basics and Advanced Methods

In the case of low volume production, where investment in more sophisticated dewatering equipment is not required or desired, there are a few different techniques that can be used to get the job done.

The three most popular methods for evaporating water from Zeoform Micro Pulp include using a solar dryer (passive or active), air heater exchanger, or a dehumidification system. These methods all have their strengths depending on your needs but they also each come with their own weaknesses. For example, while an inexpensive option like a window sill may work well when conditions are favorable it may produce uneven drying which may prove undesirable.

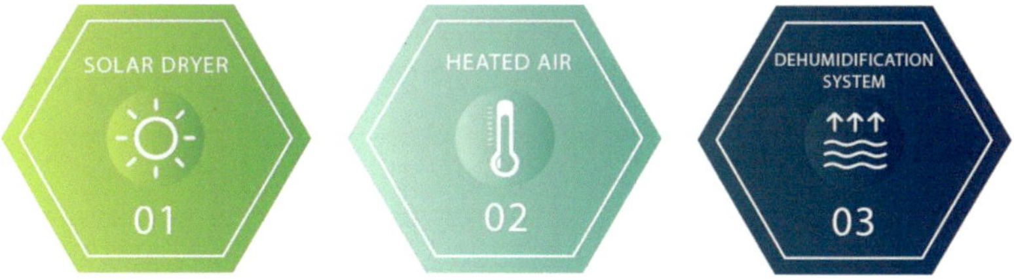

Tactic 3: Material Storage

The goal of any storage system is to make it easy for your employees and customers to find what they need. The considerations for this process are vast, but the key points should help you narrow your decision. There is no perfect storage system that will work for all products in every scenario. You need to ask yourself what's most important to you and make a choice accordingly.

If ease of access is really high on your list, opt for something like palletised shelving or warehouse rack systems with rollers so items can be easily moved around when needed. Savings will also depend on how many units you want to store at once; if there are only 10-20 units then picking up some self-storage shelves may save money over renting out an entire unit from a warehouse company that offers both storage space and equipment rental together as one package.

Tactic 4: Material Stock Control

There are many best practices for stock control. It is important that you know all of these in order to run your business smoothly. Material stock control can be a difficult topic if not properly managed and has been known to wreak havoc on businesses who don't take the time necessary to prepare before opening their doors or launching a product marketing campaign.

It's worth noting that there are other considerations as well, such as what material price fluctuations mean and how to make sure you're always meeting customer demand at a competitive cost point.

Tactic 5: Material Shipping

Zeoform Micro Pulp is a product that can be shipped all over the world. It's important for you to ensure your customers receive their order in perfect condition and on time. So make sure every detail of the shipping process matters. You will need to consider everything from packaging to label placement as well as other things that can impact your package's journey from factory floor to customer doorstep. You will also need to consider what sort of transportation methods are available, and which carriers to use.

Tactic 6: Material Quality Assurance

Quality assurance is the process of ensuring that a product or service meets certain standards. It is important for all businesses to have quality assurance in place before releasing their products because it ensures that everything about the final product will meet expectations before going out to consumers.

Without quality assurance procedures there can be problems with products which impact customer satisfaction and brand trust, not to mention waste money on unnecessary fixes after release.

Tactic 7: The Production Innovation Process

If you're a leader of an organisation, formal or informal, it's time to adapt. The world around us is changing rapidly and those who have the ability to innovate will be best positioned for success in this new environment. Innovation as a core competency should be at the forefront of your mind if you want sustainable outcomes from your business strategy.

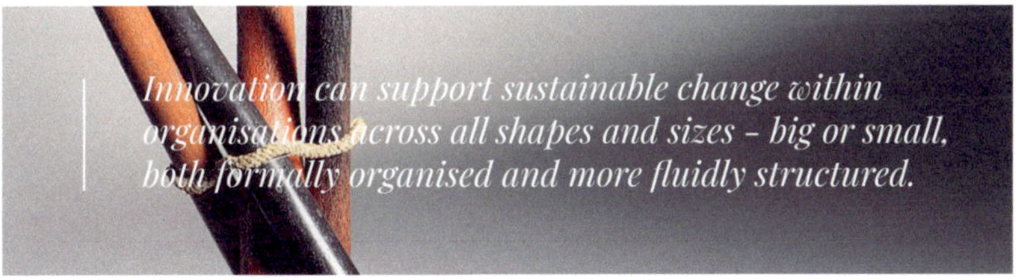

While innovation is still a relatively young discipline, it has become clear that all organisations need to build their workforce capacity for innovation. The product innovation process can help you do this by identifying your organisation's strengths and weaknesses in order to create new products with the best chance of success. The product innovation process is a critical component of the success or failure of any business.

Each of the 5 steps in the path to innovation is a distinct step and requires specific attention. At each step there exists a risk of failure. However, the biggest risk to achieving effective results is facilitating acceptance and buy-in of the solutions developed.

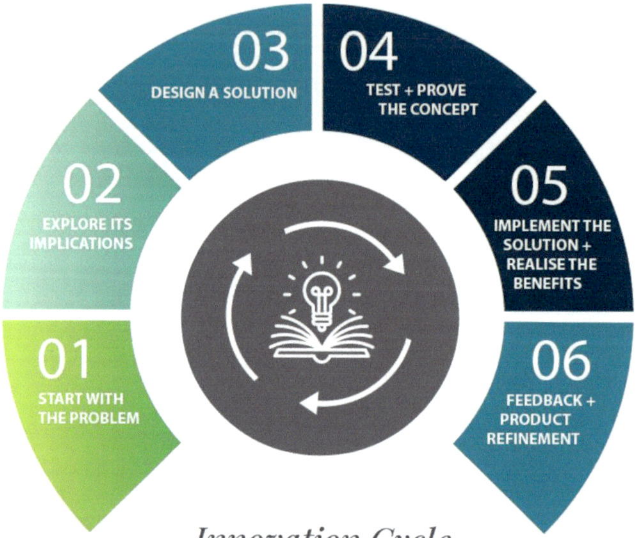

Innovation Cycle

Innovation is the creation of an effective new result in that something is improved or its value has increased. This can include how your business generates revenue and profit, develops new products and services, serves customers, staff, shareholders, society and the environment.

- Articulate the issues, needs and opportunities for your future product.
- Understand the costs, implications and strength of desire to solve the problem.
- Bridge the problem to a matching solution that incorporates constraints and opportunities.
- Start the research and development phase.
- Design with purpose and understand your market for the best possible outcome.
- Work with your clients, customers and consultants to optimise your redesign process.

Tactic 8: The Future of the Construction and Furniture Industry

The effects of demographics and consumer behaviour changes on the environment are expected to be significant. Material innovation and product evolution will play a key role in mitigating these environmental impacts, as well as creating new opportunities for growth.

The construction and all adjacent industries need to adapt during these changing times. Material innovation and product evolution are the keys to success for any business in this competitive landscape.

A lot of this has to do with population growth; it's projected that there will be 7 billion people living in cities by 2050 according to the European Commission, which means increased urbanisation and waste accumulation. Rapid urbanisation leads to an increased demand for infrastructure like housing, transport, road, education and health care services. With material innovation and product evolution, companies can meet these needs now instead of waiting until they become urgent problems down the line.

We emphasise the use of biobased materials, which are sourced preferably from waste materials like agricultural bi-products, amongst others. The construction and furniture industry can also benefit from using materials that are not just sustainable, but also lighter in weight. Lighter materials can indirectly decrease environmental impacts through lower energy consumption in use, for example, lighter cars use less fuel. Companies should be able to make their products from a fraction of the natural resources they do today for this reason.

Buildings and construction are the #1 cause of the climate crisis. To solve the climate crisis, we need to solve the challenge of emissions from buildings and construction. These emissions come not just from heating and cooling, but also from the carbon cost of materials used in construction.

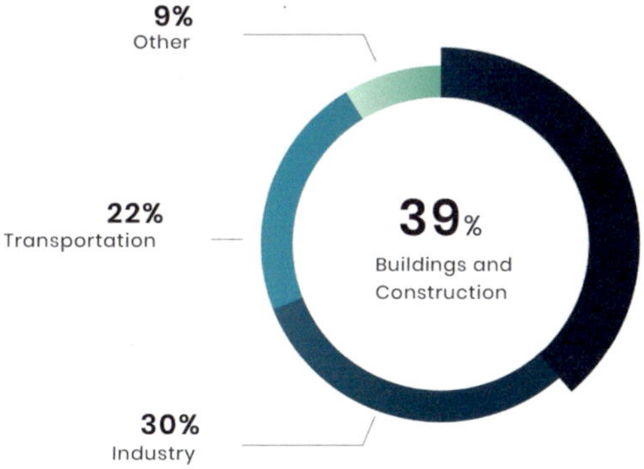

Tactic 9: New Developments in Wood Replacements, and other Adjacent Markets

It's been a few years since we last saw transformative building materials introduced to the construction industry, so it should come as no surprise that sustainability advocates and other stakeholders are a buzz over Zeoform Micro Pulp.

NMFC has been recognised as a material that will redefine the future and as being at peak commercial readiness. Architects, builders, and academic researchers are already interested in this innovative product and see it becoming a major industry opportunity for those that wish to contribute to the circular economy by upcycling waste streams and using carbon-sequestering technologies.

Figure: Zeoform Micro Pulp Sustainable Drivers of Change

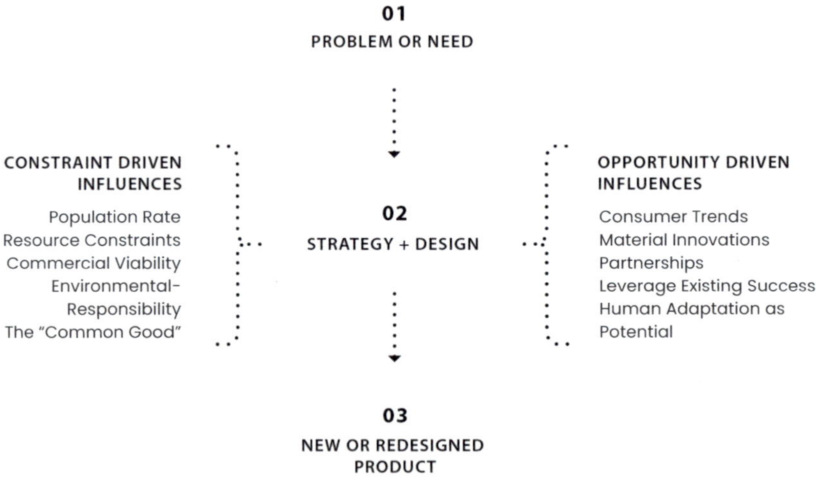

Tactic 10: The Research + Development Process

One of the most daunting, but necessary steps in launching your own product is coming up with an idea for what you want to produce. The process can seem almost mysterious and when you hear about how other great products came into fruition, it's rare that their journey resembles a straight line.

It may help to know some tips on getting started with product development before diving headfirst into the unknown waters of inventing something new. Production innovation has helped many companies come up with unique ideas or improve existing designs by following proven steps like design, prototype development, production sourcing and more.

The Product Research and Development Process

1. **Idea generation** - ideation, iteration and strategy assessment.
2. **Research** - product validation from an unbiased, independent audience.
3. **Planning** - turning a concept into a detailed design.
4. **Prototyping** - creating a minimum viable product and variations for testing.
5. **Sourcing** - production planning for materials, partnerships and expertise.
6. **Costing and viability study** - financial and other performance analysis.
7. **Launch and proof of concept** - testing uptake to reach critical mass.

Summary + Knowledge Checklist

Factors for Success

Tactic 1	Water evaporation fundamentals
Tactic 2	Advanced methods for water evaporation
Tactic 3	Material storage fundamentals
Tactic 4	Understanding material stock control
Tactic 5	Understanding material shipping basics and considerations
Tactic 6	Understanding material quality assurance and product delivery consistencies
Tactic 7	The production innovation process
Tactic 8	The future of the construction and furniture markets
Tactic 9	Developments in wood replacements, and other adjacent markets
Tactic 10	The research and development process
Tactic 11	Understanding refiner cycles and passes
Tactic 12	Gaining knowledge of energy consumption
Tactic 13	Learning about fibre & water ratios
Tactic 14	Key variables for how to empower the active production process

ZEOFORM™

Zeoform Transformation Applications

Overview

Zeoform Value Transformation

Method Applications

Chapter Overview

Zeoform Value Transformation is a Method for enabling sustainable manufacturing. It provides a results-focused framework for helping business owners to realise their sustainability objectives quickly, easily and confidently. Applications for the Method include:

- A blueprint for a start-up business model.
- A management tool for sustainable management in teams to assess the readiness in their sustainable production processes.
- A coaching tool for managers when exploring new avenues for sustainable manufacturing.

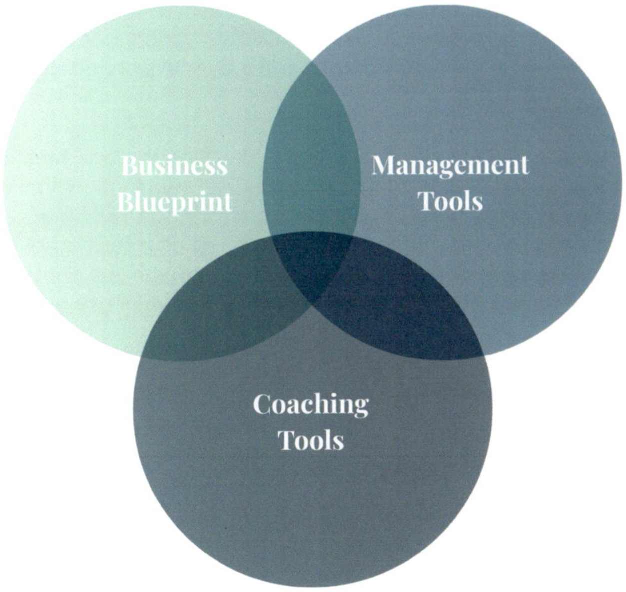

Zeoform Value Transformation

Method Applications

Zeoform Value Transformation as a blueprint for a start-up business Method

Small to mid-sized businesses are under a lot of pressure right now to evolve and position themselves for the new world of business. The post-pandemic world has also added additional constraints (and opportunities) for small to mid-sized businesses to move quickly into new supply chains and markets.

Whether these businesses are based in Australasia, North America, Europe, Asia or anywhere else in the world, businesses are feeling the impetus to evolve quickly to the needs of a sustainable future.

As with many new developments each business needs an X Factor that gives them the competitive advantage and helps to de-risk the movement into a new strategic direction. When expert leadership is absent, indeed expertise in the chosen area, the risks of undertaking a new strategy increases.

In the context of the Zeoform Value Transformation Method, businesses are given well structured and highly researched information so they can plot a clear road map into the future. From a management point of view using the Zeoform Value Transformation Method as a starting blueprint can help accelerate your business to step through the key decision points needed to assess the value of a new opportunity using Zeoform Micro Pulp as an innovative raw material.

Zeoform specialist consultants are available to assist you every step along the way.

Zeoform Value Transformation as a Management Tool for Sustainable Businesses

Project teams that are applying sustainable management to their businesses will usually prepare plans and strategies and other management documentation. The Zeoform Value Transformation Method can be used as a checklist so you can evaluate the completeness of these projects. For example, assume that you have decided you need to enter the world of sustainable manufacturing. Your new strategy and roadmap should include an assessment of the pre-process, the active process and the post active process in order to achieve your key business objective and sustainable goals.

If you use the Zeoform Value Transformation Method as a management tool you can work quickly towards identifying the required knowledge, equipment and expertise you need to build an effective business model.

Zeoform Value Transformation as a Coaching Tool for teams assessing their sustainable manufacturing projects

Managers and business development specialists can use the Zeoform Value Transformation Method as a coaching tool with their staff and teams during strategic initiatives. Consider the many elements you will need to think about when developing new business ideas and Methods to grow (and protect) the future of your business.

The Zeoform Value Transformation Method can help you structure and identify the key elements of a sustainable manufacturing business plan and therefore positions you to make better decisions about entering the renewable materials market. If you have any need for additional assistance or support from Experts in the field please do not hesitate to contact Zeo IP.

ZEOFORM™

Zeoform Value Transformation Review

Overview

Zeoform Value Transformation Overview

9 Step Zeoform Value Transformation Method Overview

The previous three (3) chapters presented a range of tactics that can be used to achieve effective results across each of the 3 stages (pre, active and post) of the Zeoform Value Transformation Method. This method provides a results-focused framework for sustainable manufacturing activities. A summary of these tactics is provided in Appendix A.

The 9-Step Zeoform Value Transformation Method is designed to engage readers from different levels of learning. Each stage of the method has a series of stages and lessons that explain the why, what, and how to activate the Zeoform Micro Pulp production method from end to end.

The Method weaves together basic, intermediate and expert level knowledge ranging from what is cellulose, how to build an innovation lab through to the jewel in the Zeoform crown - how to make Zeoform Micro Pulp with distinct fibre properties.

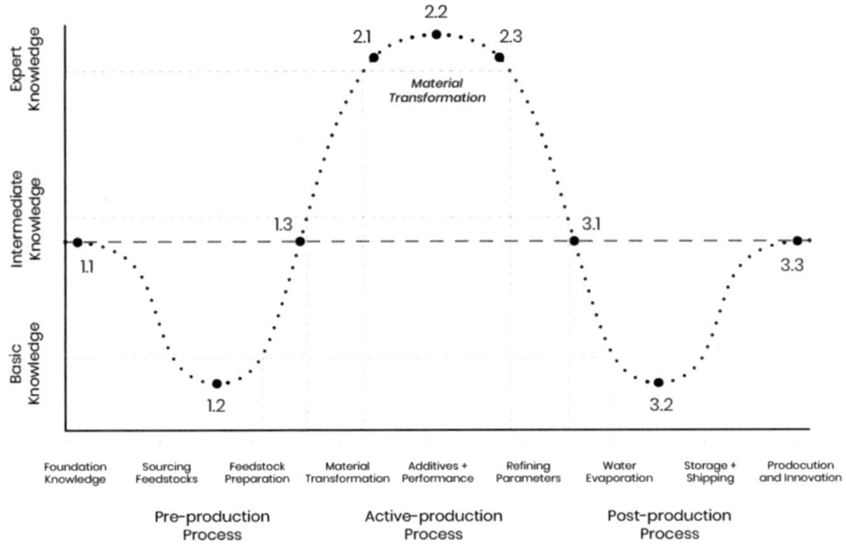

Stages	Steps	Tactics (Lessons)
Stage 1 Pre-production process	Foundation knowledge	Gain a deeper knowledge of cellulose Understand more about plant fibre morphology Explore the future of manufacturing Awareness of methods for nano and microfibrillated cellulose (NMFC) Understand the basis of mechanical refining to make NMFC
	Sourcing feedstocks	Discover the abundance of cellulosic biomass Identify commonly available feedstocks in your area Explore all considerations that make a feedstock viable Explore globally available feedstock such as hemp, bagasse and bamboo Understand the importance of identifying feedstock substitutes
	Feedstock preparation	Gain knowledge of feedstock planning Feedstock volumes and quality Look at preparation methods such as washing and pre-pulping Identify opportunities for material performance enhancements Look at other factors that affect the production process
Stage 2 Active-production process	Material transformation	The Magic of Cellulose 9 Step Zeoform Value Transformation Method Stages in cellulose transformation Benefits of Using Zeoform Micro Pulp
	Additives & performance	Material characterisation Material additives Pigments and Dyes Materials applications
	Refining parameters	Researching mechanical refiner types Researching refining blade setups Understanding refiner cycles and passes Gaining knowledge of energy consumption Learning about fibre & water ratios Other variables
Stage 3 Post-production process	Water evaporation	Water evaporation fundamentals Water evaporation basic methods Water evaporation advanced methods Water evaporation other variables
	Storage & shipping	Material storage Material stock control Material shipping Material quality assurance Other storage and shipping variables
	Production innovation	The innovation process Developments in construction industry Developments in furniture industry Developments in adjacent and other markets Material application research and development

Next Steps

Conclusion: Transformation Method

Zeoform Value Transformation as a Management Tool for Sustainable Businesses

The Zeoform Value Transformation Method has three stages that must be achieved in order for sustainable manufacturing to be implemented and sustained. These stages form the basic building blocks for successful, sustainable manufacturing:

1. **Stage 1:** Pre-production process
   Planning and readiness
2. **Stage 2:** Active-production process
   Manufacturing process and knowledge
3. **Stage 3:** Post-production process
   Product distribution

After 10 years of developing this method and applying this method with business, government representatives, and community interest groups, we have observed the "light bulb" and the excitement coming on many times. The simple method of utilising the most abundant source of renewable raw material on the planet and how to transform it into a highly versatile game-changing raw material enables business managers worldwide to connect quickly and easily with sustainable manufacturing practice.

Once the key concepts are understood and you've implemented them through your strategic planning process a couple of times you will awaken to the new world of refining cellulose for advanced manufacturing. I'm sure that right now you're feeling like you've been drinking from a firehose, and we totally understand that. The concepts included in this book took us over 10 years to discover from dozens of different experts and through a whole lot of trial and error. We wish we'd had something like this when we first started on our journey. We hope that instead of feeling overwhelmed you realise that this is actually just a huge shortcut to a new world of good business - in this new world there are three key elements worth mentioning.

The goal of the Zeoform Sustainability Formula is to make a key contribution to a thriving world. In this world businesses are developing products and services that are commercially viable, environmentally-responsible, and contribute positively to the common good.

As we noted earlier in this book:

Commercial Viability represents an understanding of the nature of the business opportunity, why the opportunity is arising and the risk of not taking up the opportunity. Commercial viability also includes information about the internal and external drivers that created the opportunity, as well as the definition of returns on investment and effort.

Environmental Responsibility represents the physical dimensions and natural environment impacts of activities through the opportunity. Responsibility means taking care and ownership for all things, including for all pre-production processes, active process and post-production activities.

Common Good represents the human dimensions of all connections and relationships between people and groups involved. Community contribution includes the levels of cooperation, collaboration and coordination required, along with support for human rights and higher states of human relationships and development.

The Zeoform Value Transformation Method provides a solid foundation for sustainable manufacturing, including outlining all key elements of the three stages of production: **pre-production process, active-production process and post-production process activities**.

If you're wondering where you should start, this is what we recommend:

1. **Become very clear about your sustainable development goals** and the new opportunities this will create on all levels of business and humanity.

2. **Engage strategically** with a diverse variety of thinkers, because it is through diversified viewpoints that new breakthroughs in thinking are created.

3. **Become your own greatest hero.** This is the most important skill you can learn. When you feel strongly about a cause that is greater than you or you have a vision for a better future, then your commitment and follow through will not just encourage you to be the best you can be but it also encourages others to step up and make a strong and positive contribution to a thriving world.

4. **Change the world.** Your efforts and commitment have the ability to transform people's lives. By committing to a sustainable future you're not only assisting the current generation but you assist the starting point for all future generations to come.

Source: https://wdo.org/about/vision-mission/un-sdgs/

Conclusion

We hope that reading this book has been a great investment of your time. We spent over 18 months writing and rewriting it because we wanted it to be perfect for you. Here at Zeo IP, we believe your business has the opportunity to provide economic and social good while developing innovative and truly sustainable solutions for all people and the planet.

If you'd like to work more closely with Zeo IP we invite you to make contact with us via our website - *www.zeoform.com*. There you will find further information and you can register your interest and someone on our team will contact you. If you are a good fit we could be working on your new business opportunity within weeks. We are big believers that success follows effort - don't let this opportunity pass you by.

With that said we encourage you to master the skills in this book and become the vehicle for change that we know your business can be. After you've done that we encourage you to go even further. Spend more time working on that and let's figure out together how to take your company, your vision, and customers into a future of sustainability and success.

ZEOFORM™

Appendix A

The Zeoform Value Transformation Checklist

Stage 1: Zeoform Micro Pulp Pre-Production Process
Areas of Key Knowledge

- ❏ Cellulose fundamentals.
- ❏ Plant fibre morphology.
- ❏ Industry 4.0 and the future of manufacturing.
- ❏ Methods for making nano and microfibrillated cellulose (NMFC).
- ❏ Mechanical refining basics.
- ❏ The abundance of cellulose.
- ❏ Commonly available cellulosic feedstocks.
- ❏ Assessing feedstock viability.
- ❏ The importance of feedstock planning.
- ❏ Feedstock volumes and quality.
- ❏ Feedstock preparation methods.
- ❏ Feedstock substitution.
- ❏ Opportunities for material performance enhancement.

Stage 2: Zeoform Micro Pulp Active-Production Process
Areas of Key Knowledge

- ❏ Cellulose in more depth.
- ❏ Nature's transformation process.
- ❏ The stages of cellulose transformation.
- ❏ Material characterisation.
- ❏ Material additives.
- ❏ Material pigments and dyes.
- ❏ Material applications.
- ❏ Mechanical refining for Zeoform Micro Pulp.
- ❏ Disc refining and blade setups.
- ❏ Understanding refining cycles and processes.
- ❏ Energy consumption efficiency and analysis.
- ❏ Plant fibre & water consistencies for refining.

Stage 3: Zeoform Micro Pulp Production Post Active-Process Areas of Key Knowledge

- ❑ Water evaporation fundamentals.
- ❑ Water evaporation basic and advanced methods.
- ❑ Zeoform Micro Pulp storage.
- ❑ Zeoform Micro Pulp stock control.
- ❑ Zeoform Micro Pulp shipping.
- ❑ Zeoform Micro Pulp quality assurance.
- ❑ The product innovation process.
- ❑ The future of construction and furniture.
- ❑ Developments in wood replacements, and other adjacent markets.
- ❑ The research and development process.

Appendix B

The Zeoform Sustainability Readiness Self-Assessment

Please complete the following exercises:

> **STEP 1:** Grade yourself against the following questions. These questions are based on the key factors driving sustainable success that are covered in *Chapter 2.*
>
> **STEP 2.** Total your scores to all five questions, and write it in the box below.

1) You're fully informed of the macro/micro trends affecting business and community

(Least) **1** **2** **3** **4** **5** *(Most)*

2) Your strategic growth and development plans are aligned with the common good i.e. global sustainable development goals and the universal rights for humans.

(Least) **1** **2** **3** **4** **5** *(Most)*

3) You're fully informed of the impact of fossil fuels consumption and the current state of waste accumulation and degradation in all ecosystems globally.

(Least) **1** **2** **3** **4** **5** *(Most)*

4) You're informed of the full range of renewable material alternatives across the full spectrum of your supply chain.

(Least)	**1**	**2**	**3**	**4**	**5**	*(Most)*

5) You are highly motivated and driven to make a positive contribution to the lives of others and, in the process, create a competitive advantage for your business.

(Least)	**1**	**2**	**3**	**4**	**5**	*(Most)*

Sum of Scores (out of 25 total)

(0) *(25)*

Appendix C

Conclusion: Factors for Success

The Zeoform Sustainability Readiness Self-Assessment

Please complete the following exercises:

> **STEP 3.** Use the following table to complete your readiness self-assessment. This assessment is based on the normal distribution model.

Overall Score	Scale	Implication
5	Not ready	You're at major risk; you're walking blind into your future.
6-12	Minimally ready	You need to invest more into your state of readiness.
13-21	Moderately ready	This is encouraging; double down on your effort right now.
21-24	Mostly ready	Leverage your readiness now to create a competitive advantage.
25	Fully ready	Congratulations! You are fully ready to embrace the new world.

The Zeoform Sustainability Readiness Self-Assessment

Please use the following:

Notes. Use the following table to brainstorm ideas and develop a successful roadmap

Zeoform Value Transformation

Please Contact our Support Team with Questions!

E. support@zeoform.com

Sincerely,
The Zeoform Team

Made in the USA
Las Vegas, NV
04 November 2026

57424373R10067